Introductory Radio
and Television Electronics

THE
SERIES

Macmillan Texts for Industrial Vocational and Technical Education

Introductory Radio and Television Electronics

James J Johnson

Brian D Fletcher

MACMILLAN

First published 1997 by
MACMILLAN EDUCATION LTD
London and Basingstoke
Companies and representatives throughout the world

ISBN 0-333 61656-1

10	9	8	7	6	5	4	3	2	1
06	05	04	03	02	01	00	99	98	97

Printed in Hong Kong

A catalogue record for this book is available from the
British Library.

Acknowledgements
The authors and publishers wish to acknowledge, with thanks, the
following photographic sources:
Steve Cartwright

Cover photograph courtesy of Telegraph Colour Library.
Illustrations by 1-11 Line Art

Dedication
This book is dedicated to our immediate families and children.

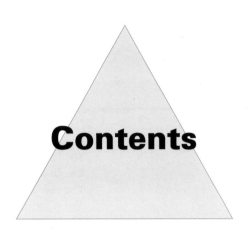

Contents

Acknowledgements

The authors would like to thank their immediate families for their forbearance while this book was being written.

Special thanks go to Rosa Crivaro, who patiently typed all James's manuscripts, and to his past colleagues Albert and Cyril. We should both like to thank Pam for her help and support.

Thanks also go to the following people and organisations, whose help has been invaluable in writing some sections of this book: Andy Sennitt, editor of the *World Radio and Television Handbook*, Amsterdam; Philips Components Ltd; The Slide Centre, London; Mr Geaves of the Ferguson Company, for allowing us to use some of their circuit diagrams; Alan Lafferty, Manager, and John Pinninger, Engineering Information, at the British Broadcasting Corporation; Steve Cartwright for the splendid photographs.

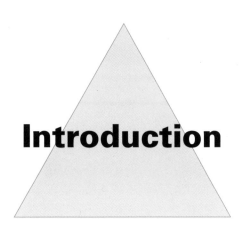

Introduction

This book provides an introduction to the principles used in the transmission and reception of radio and television signals. The book's scope assumes a working knowledge of basic electronics; it therefore does not cover subjects such as audio amplifiers.

The book should appeal to students studying for examinations in radio and television, and those taking options in these subjects within other examinations. It should also be helpful to service technicians who wish to refresh their understanding of the subject.

The text has been written to be easy to follow. The use of complex mathematics has been kept to a minimum; the emphasis is on the practical and descriptive aspects of the subject rather than on the theory. Detailed diagrams and photographs have been included to assist in learning and understanding the subject. Many circuits have been reproduced from manufacturers' service manuals. At the end of each chapter, there is a 'Check your understanding' section followed by self-test and proven exercise questions. These should assist in reinforcing the reader's knowledge and understanding, and allow for self-checking.

The book is introductory, and so we have concentrated on the most important aspects of the subject. Some topics are explored in detail; others are discussed in outline, the reader being left to follow up these areas using other sources. Because of the amount of information covered within the book, and the nature of this special subject, this text can also be used as a basic reference book.

The nature of radio waves

Introduction

This chapter establishes some basic facts about the nature of radio waves. We first look at the way in which any wave can be made up from basic sine waves. We then go on to the terminology that is used in describing sine waves, to the frequency spectrum, and to the characteristics of different parts of the frequency spectrum. Finally we look at the electromagnetic wave, and the model that has been constructed to explain its existence.

The basic sine wave

All types of waveform are derived from a basic wave called a **sine wave**.

If we look at the sine function using a calculator or mathematical tables, we can deduce the pattern of the sine wave by plotting sine ϕ against ϕ for angles between 0 and 360°. If you are not familiar with the shape of the sine wave, try plotting it on graph paper.

Another way of showing a sine wave is by using a **rotating vector** – an arrowhead rotating anticlockwise in a circle of 360° from a centre point. The length of the arrowhead will indicate the maximum positive and negative values of the sine wave. You can draw the sine wave (Figure 1.1) by connecting the arrowhead of the rotating vector horizontally, for various angles, to intercept with the vertical lines projected from the same angles on the angles axis.

When vectors are used to represent electrical sine waves they are usually referred to as **phasors**.

We need to examine the characteristics of a sine wave so that it can be described in words. The **amplitude** is the maximum value of the excursion of the wave in a positive or negative direction. The **frequency**, expressed

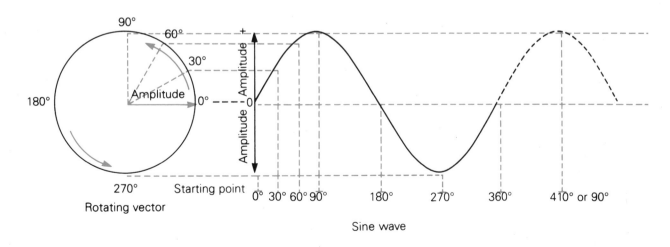

Figure 1.1 Rotating vector method of showing a sine wave.

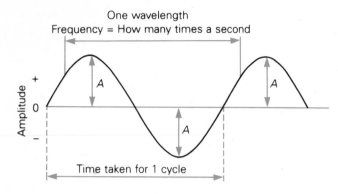

Figure 1.2 Amplitude and frequency of a sine wave.

as the symbol *f*, is the number of times the wave completes a cycle of values in one second: that is, the number of times that it goes from 0° through 360° back to 0°, or likewise completes 360° from any point in its cycle.

Frequency is measured in units called **hertz** (abbreviation Hz), which means **cycles per second**. A sine wave that completes 10 complete cycles in one second has a frequency *f* of 10 Hz; one that completes 100 000 cycles in one second has a frequency of 100 kHz.

Note that the abbreviation for 1000 is kilo (k). The abbreviation for 1 000 000 is mega (M), and for 1 000 000 000 is giga (G). See other abbreviations in Appendix 4 at the end of the book.

Figure 1.2 illustrates the meaning of amplitude and frequency.

Another useful word in radio and TV associated with the sine wave is the term **harmonic**. This refers to a multiple of the frequency of a **fundamental** sine wave (the original sine wave). That is, if the fundamental sine wave has a frequency of 100 Hz, then the harmonics of this sine wave will be at frequencies that are multiples of 100 Hz: 200 Hz, 300 Hz, 400 Hz, 500 Hz etc. These are called the second, third, fourth and fifth harmonics respectively. For obvious reasons the second and fourth harmonics are called **even harmonics**, and the third and fifth harmonics are called **odd harmonics**.

> Any waveform can be considered as being made up of a number of sine waves of different amplitudes and frequencies.

This makes the sine wave a very powerful tool in analysing waveforms. It also gives us a model for considering many aspects of radio, television, and communications.

One of the most complicated waveforms that is used

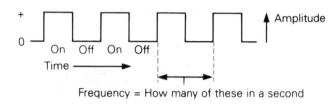

Figure 1.3 On–off waveform or square wave.

extensively in modern communications is the on–off waveform or **square wave** (Figure 1.3). This can be shown to be made up of a sine wave fundamental at the frequency of the square wave plus an infinite number of odd harmonics of that fundamental waveform.

ACTIVITIES

1   You can demonstrate the make-up of the square wave using the equipment shown in Figure 1.4. A reasonable square wave can be obtained using a fundamental plus the third, fifth, seventh and ninth harmonics, with decreasing amplitudes according to the harmonic used. For example, the third harmonic would use a third of the amplitude of the fundamental, the fifth would use a fifth of the amplitude and so on. In mathematical terms, this is using a progression in which the amplitude of the harmonic is divided by its number.

2   As an alternative, try to produce a square waveform on graph paper, by adding a sine wave fundamental and the harmonics described above.

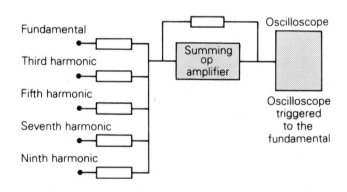

Figure 1.4 Equipment needed for Activity 1, to demonstrate the make-up of the square wave.

Wavelength

Another way of describing waves, especially radiated waves, is in the form of **wavelength**, which is given the symbol lambda (λ). Wavelength is usually expressed in metres.

$$\text{Wavelength, } \lambda = \frac{\begin{array}{c}\text{Velocity of the wave in}\\ \text{the medium it is travelling in}\end{array}}{\text{Frequency of the wave}} = \frac{v}{f}$$

where λ is expressed in metres (m), f is in Hz, and the velocity v is in metres per second (m/s). As sound waves travel through the medium of air at 330 m/s, the wavelength can easily be calculated: the higher the frequency, the shorter will be the wavelength.

Radio waves use the medium of space. They travel at the speed of light in a vacuum, which is 300 000 000 m/s (186 000 miles per second). This velocity is given the symbol c. So for radio waves, $\lambda = c/f$ m. Thus a frequency of 3 MHz corresponds to a wavelength of

$$\frac{300\,000\,000 \text{ m/s}}{3\,000\,000 \text{ Hz}} = 100 \text{ m}$$

> The basic concept is that the higher the frequency is, the lower or shorter will be the wavelength, and vice versa.

The use of wavelength when considering radio and communication has long been established. Many older radio receivers have tuning dials marked in wavelength, but today it is common for tuning dials to be marked in frequency.

The wavelength or frequency spectrum

Sound waves

These waves are generally regarded as being those below 20 kHz. They are transmitted using the medium of air or liquid by causing the medium to be compressed and rarefied alternately. Just how quickly this is done represents the frequency, while the violence that accompanies it represents the amplitude.

We can obtain a useful analogy by disturbing a pool of water. If we disturb the water by throwing a stone into it, a wave will travel outwards from this disturbance. If we throw more stones in, the number thrown in one after another will indicate the frequency of the waveform, while the size of the stone will be an indication of the amplitude or size of the disturbance that travels across the pool.

As we have already seen, sound waves travel at 330 m/s at sea level, which is much slower than the speed of light, at 300 000 000 m/s. This is why, when the sound of an event travels a long distance, there is a delay between the eye seeing the event and the ear hearing it. Perhaps the most spectacular example of this is in a thunderstorm, when a flash of lightning may be seen long before the crash of thunder is heard. The time difference will depend on the distance between the listener and the point of the lightning flash, and therefore it is possible to work out roughly how far away the lightning is.

As humans, we use sound waves as our main means of communication. The range of frequency used for the spoken word is approximately 0–5000 Hz. The human ear is capable of hearing frequencies from 25 Hz up to about 15–20 kHz, depending on age; the range falls off with age. The fundamental frequency of the human voice is considerably lower than the range quoted; it is the harmonics that make up the range and also the characteristics of the different voices that humans have and the sounds they make. For reasons that will become apparent later, any communication of voice (such as telephone or radiotelephone) has a restricted range of 300 to approximately 3500 Hz. This means that on the telephone it may sometimes be difficult to identify a person's voice correctly, because the full range of harmonics has been restricted.

ACTIVITY

3   Connect an audio frequency (AF) generator and a loudspeaker (or earphones) as shown in Figure 1.5. Increase the frequency until it is not possible to hear the sound any more. Record this frequency. Repeat the exercise for people of different sex and ages. Is there any correlation?

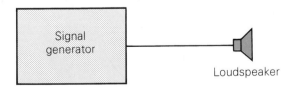

Figure 1.5 Equipment needed for Activity 3, to test people's range of hearing.

Sound waves can be heard only at a finite distance, and they can become disturbed if many sounds are taking place at the same time: think of a crowded room with everyone talking together. This makes it difficult to try to communicate with someone over a long distance using sound waves, and therefore some medium must be found to transmit the sound waves. One way of doing this is to convert the sound waves by means of a **transducer** (microphone) into electrical voltage variations, and then send these variations over distances using physical wire lines to a transducer (earpiece) at the receiving end, where they can be converted back into sound waves (a telephone). However, this method has the restriction that, without further sophistication, only one conversation can take place at any one time.

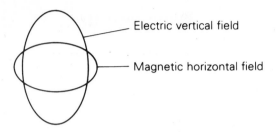

Figure 1.7 The electromagnetic wave.

component of the wave. That is, if the electric component is vertical the polarisation is said to be vertical. This is an important point when considering communications and the transmission and reception of radio waves using aerials.

Radio waves

Radio waves are part of the **electromagnetic wave spectrum**, which starts at low frequencies and goes up to extremely high frequencies, eventually becoming first light and then X-rays (Figure 1.6). For communication purposes we are concerned here with the range of the spectrum below the infrared.

Radio waves travel through the medium of space by a process that is not completely understood. Because of this a model has been developed that satisfies the current knowledge requirements. It is called the **electromagnetic wave**. The model consists of an electric wave and a magnetic wave, 90° apart, which cannot exist without each other (Figure 1.7). The two inseparable components of this electromagnetic wave travel together through free space at 300 000 000 m/s (the same as light). This speed reduces slightly if the wave is travelling through any other medium.

The **polarisation** of the wave is important in communications, and is governed by the electric

Radio frequencies

The range of frequencies that make up the **radio range** has gradually become wider, in the sense that radio transmissions now extend into the VHF, UHF, SHF and microwave ranges; these last are used for satellite and point-to-point (from one point to another, usually in a direct line of sight) communications. This is a matter of general technical development in terms of equipment, of ideas, and most of all of the space available on the spectrum for communication. Figure 1.8 shows the radio communication frequency spectrum with some of the bands that are used for specialist services.

■ **CHECK YOUR UNDERSTANDING**

● All waveforms can be derived from the sine wave and its harmonics.
● Frequency, symbol f, is the number of times that a waveform completes its complete cycle in one second. It is

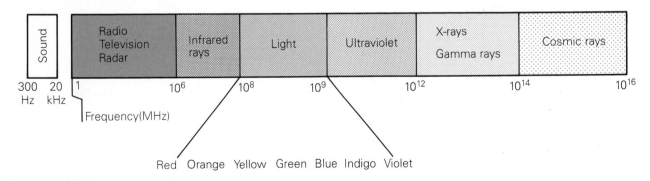

Figure 1.6 The electromagnetic wave spectrum.

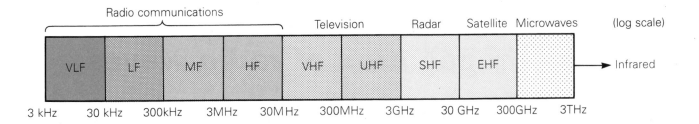

Figure 1.8 The radio communication spectrum. VLF, very low frequency; LF, low frequency; MF, medium frequency; HF, high frequency; VHF, very high frequency; UHF, ultra high frequency; SHF, super high frequency; EHF, extremely high frequency.

measured in cycles per second or hertz (Hz).

● Amplitude is the maximum excursion of the wave in a positive or negative direction from the norm.

● Harmonics of a waveform are multiples of the fundamental frequency of the waveform.

● Wavelength in metres is equal to the velocity of the wave in the medium it is travelling in (metres per second) divided by the frequency of the wave in Hz.

● The human sound bandwidth is from approximately 25 Hz to 15 kHz.

● Telephone frequencies are generally restricted to the range 300–3500 Hz.

● The radio or electromagnetic wave is modelled in terms of inseparable electric and magnetic components displaced by 90°. The electric component decides the polarisation of the wave.

● An electromagnetic wave travels in free space at 300 000 000 m/s.

● The radio wave spectrum is from approximately 10 kHz to approximately 1000 MHz.

REVISION EXERCISES AND QUESTIONS

1 A sixth harmonic sine wave has a frequency of 3000 Hz. What is its fundamental frequency?

2 Explain what you understand by the terms 'frequency' and 'amplitude' when applied to a sine wave.

3 Calculate the frequency of an electromagnetic wave when travelling in free space if it has a wavelength of: (a) 3 cm, (b) 1000 m, (c) 200 m.

4 Calculate the frequency of a sound wave just above sea level if its wavelength is: (a) 3.3 m, (b) 13 cm.

5 Which part of the electromagnetic frequency spectrum is used for point-to-point communications and satellite links?

6 Draw the model of an electromagnetic wave with horizontal polarisation.

7 Using the equipment in Activity 1, connected as in Figure 1.4, feed in different values of sine waves to obtain different waveforms. Record your results.

8 Record different ascending notes from a musical instrument such as a piano or a recorder. Using a frequency spectrum analyser, check the frequencies of these notes. Compare the results by using a storage oscilloscope to check the frequency. Note that each note has a fundamental frequency and harmonics, which will make it difficult to record the fundamental frequency.

9 Find a suitable location where you can observe a position that is at least 1 mile (1.6 km) away. Get someone to fire a starting gun blank at that position. Note how long it takes between observing the smoke from the gun and hearing the noise of the gun firing. This illustrates the difference between the speed of light and the speed of sound.

The propagation of radio waves

Introduction

The knowledge that electromagnetic (EM) waves could be used for propagation of radio waves came as a giant step forward for long-range communications. It has meant that worldwide communications have been possible using the various propagation methods described in this chapter. Before satellites were launched the main type of worldwide communication used the ionosphere as a refracting medium for radio waves. However, with the advent of satellite communications, the main bulk of reliable communications traffic has moved to this medium. The expansion of television services has also meant that advantage has been taken of satellite relays to reach a wider audience than was possible with 'line of sight' communications.

The transmission and reception of EM waves

At frequencies at and above approximately 10 kHz, EM waves can be created by transmission lines that have electric current flowing and which are deliberately not terminated correctly (that is, are left open-ended). When this happens the power going into the transmission line is not all returned; the balance is radiated in the form of radio or EM waves. The amount of power radiated in this way will generally be a function of frequency and of the length of the transmission line.

Transmission or radiation of EM waves takes place from **transmitting aerials**. The length of the aerial (antenna) is directly related to the wavelength of the EM wave; for practical reasons it is usually a quarter or half wavelength.

ACTIVITY

1 Work out the length of aerial wire required for transmitting an EM wave of frequency 200 kHz using a quarter-wavelength aerial.

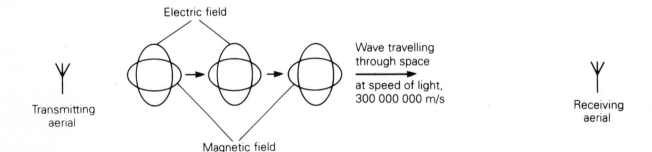

Figure 2.1 Transmission and reception of EM waves.

Radio or EM waves have the property of inducing voltages in any wire that they may pass, if the polarisation of the wave and the wire is the same. If the wire is then connected at one end to earth (ground), current will flow, and can be used to sample the EM wave that is being received by the wire. This receiving wire is referred to as a **receiving aerial**. The amount of e.m.f. induced will again depend on the length of the aerial in relationship to the wavelength and the strength of the EM wave at the time (referred to as **signal strength**). Figure 2.1 shows the transmission and reception of EM waves.

In free space (that is, a vacuum) the EM wave suffers no attenuation. In other media it is attenuated, depending on the actual medium it passes through. In many ways the EM wave behaves very much like light: it can be absorbed, reflected and refracted, just as light can.

Although no absorption of the EM wave takes place in free space there can still be a considerable reduction in the EM wave strength at the receiving aerial, because the signal strength is reduced by a factor proportional to the square of the distance from the transmitting aerial.

Consider a fictitious **isotropic radiator** (a point radiator in space – see Figure 2.2). The EM wave will travel out in all directions from the isotropic radiator in the form of a sphere. If we consider a point on the sphere at radius r metres, and the power radiated from the point source is P watts, then the power P is distributed over the area of the sphere, which is $4\pi r^2$. So the power received at the radius r, P_d, will be

$$P_d = \frac{P}{4\pi r^2} \ (\text{W/m}^2)$$

So the signal strength will decrease in direct proportion to the inverse of the square of the distance from the aerial.

This is sometimes referred to as the **inverse square law**. It also explains why special directional aerials are needed so that the amount of power required to be fed into a transmitter aerial is not disproportionately large.

Modes of EM propagation

Radiated energy in the form of EM or radio waves will travel from the transmitter aerial to the receiver aerial by one or more of three different modes:

1. surface waves or ground waves;
2. sky waves;
3. space or direct waves.

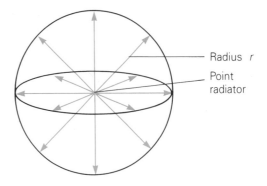

Figure 2.2 The radiation pattern of an isotropic radiator.

The last mode can be further subdivided into three: (a) direct waves, i.e. point to point; (b) via a satellite; and (c) scatter.

Surface or ground waves

This refers to the main mode of propagation for frequencies from 10 kHz to 3000 kHz. As the name indicates, in this mode the radio wave stays close to the surface or ground. It does this for a number of reasons, which vary with the wavelength of the wave.

At frequencies below 30 kHz (very low frequencies) the wavefront of the EM wave is very large. It is guided round the earth by a layer of ionised gas referred to as the **D layer**. This layer is caused by the influence of the sun's radiation at approximately 50 km above the earth's surface. In this way worldwide communication may be possible, and indeed VLF propagation was at one time used in the UK for time signals (Rugby radio on 16 kHz). It is also used for naval communications with submarines, as the wavefront tends to penetrate some way into the upper levels of the sea. Little or no absorption takes place at these frequencies in either the D layer or the surface of the earth (Figure 2.3).

At frequencies above 30 kHz and below 3 MHz (low and medium frequencies) the D layer becomes an

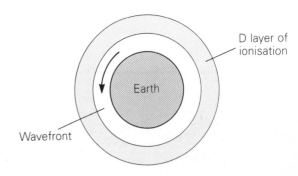

Figure 2.3 VLF propagation.

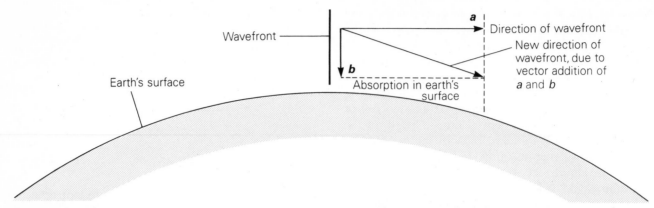

Figure 2.4 Low- and medium-frequency propagation. **a** represents the forward component of the wavefront; **b** represents the downward component of the wavefront (due to the earth's absorption).

absorption medium, attenuating any sky waves (see below). The earth's surface also absorbs energy from the EM wavefront. This absorption increases with frequency. It also depends on the nature of the earth's surface, varying from a minimum over sand or water to a maximum over dense green foliage (forests).

Because of the increased absorption of the earth's surface the wavefront becomes tilted towards it. The curvature of the earth causes the wave to follow the earth's surface closely. Absorption in the earth's surface increases with frequency, until at over approximately 3 MHz (dependent on surface terrain conditions) the surface wave plays very little part in the propagation process (Figure 2.4).

Sky wave propagation

Sky wave propagation takes place between approximately 3000 and 30 000 kHz (high frequencies, or HF). It is due to the refractive properties of the sun-activated ionised layers of gas above the earth. This gives worldwide long-range communication. The effectiveness of this communication is dependent on the frequency used, the time of day or night, the season of the year, and the sun's sunspot cycle (Figure 2.5).

The ionosphere

To explain sky wave propagation, we need to look at the ionised region that surrounds the earth. The **ionosphere** consists of four layers of ionised gas at different distances above the earth's surface. These ionised layers are caused by certain types of radiation from the sun, and so they are affected by the position of the sun in the sky. Maximum

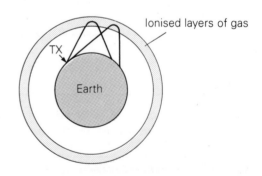

Figure 2.5 Sky wave propagation (TX, transmitter).

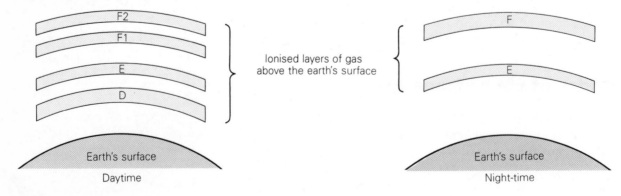

Figure 2.6 Ionised layers surrounding the earth.

ionisation occurs when the sun is overhead (midday), in summer, and when the sun is nearest the earth. It follows therefore that minimum ionisation takes place at midnight, during winter, and when the sun is furthest from the earth. The process is further complicated by sunspot activity.

Figure 2.6 shows the make-up of the ionised layers during the day and at night. During the day there are four layers, whose density is a maximum at the following approximate heights above the earth: D, 50–60 km; E, 130 km; F1, 200 km; and F2, 280 km. At night the D layer virtually disappears, while the density of the other layers decreases, and F2 merges with F1. The D layer plays no part in the actual refraction of the EM wave, but does have an attenuation effect at higher MF and lower HF. The density of each layer is a maximum in the middle, and falls off on either side (Figure 2.7).

Refraction takes place in the same way as for light. The wave is refracted away from the normal when going from less to more dense ionisation, and towards the normal when going from dense to less dense. The amount of refraction that takes place is dependent on the frequency of the wave. At low frequencies there is maximum refraction, and at high frequencies there is little refraction. At approximately 30 MHz radio waves are not refracted, and escape into space (Figure 2.8).

If the E layer does not refract the wave then it goes to F1, and possibly F2, before either being refracted back to earth or escaping into space. Each layer the wave passes through may introduce some absorption, so if a wave is refracted back through three layers some attenuation may be introduced.

Communications using the sky wave

From the brief description above we can see that the effectiveness of communications between two points using the sky wave will depend on the frequency and the time of day. Thus to maintain worldwide communications between two points it will be necessary constantly to change frequency during 24 hours as the layers of ionised gas change. See Figure 2.9, which shows the

effect of changing skip distance for a given frequency during the day.

Skip distance is the distance between the end of the ground wave and the first return of the sky wave. Typical frequencies used on a summer's day at 1400 hours to give a skip distance of 3000 miles (4800 km) would be 16 or 22 MHz, while at 0200 hours to give the same skip distance would require 8 or 12 MHz. Corresponding frequencies to give the same skip distance on a winter's day would be 12 or 16 MHz and at night would be 6 or 8 MHz.

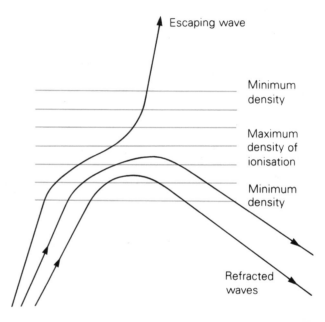

Figure 2.8 Refraction of the sky wave. If the wave is not 'refracted' by 90° by the time reaches the densest part, it will escape.

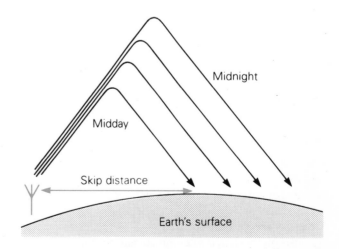

Figure 2.9 Varying skip distances.

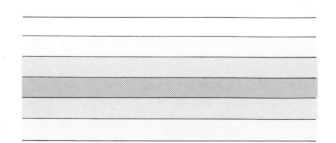

Figure 2.7 Intensity of ionised layers.

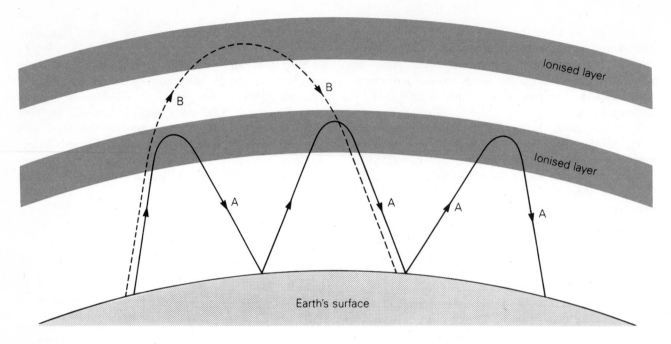

Figure 2.10 In and out of phase fading problems. Addition of wave paths A and B could be in or out of phase at point C.

The disadvantage of using sky wave propagation is the need to change frequencies to maintain communications throughout the day, seasons and years, because of the constantly changing nature of the ionosphere. Associated with this disadvantage is the problem of **fading**. This can be caused by the movement of the ionised layers, which usually occurs between the times of dusk and dawn, when the change in the sun's radiation is greatest. This movement causes the skip distance to vary, which in turn can cause the receiving aerial to be in or out of the reception zone.

Another form of fading is caused by the reception of the signal via two separate paths of refraction through the ionised layer, which can be in phase at one time and out of phase at another (Figure 2.10).

Both these types of fading can be very annoying, and can cause problems with communications. The first type can be avoided by trying not to communicate when the ionosphere is changing: that is, at dawn and dusk. The second type of fading can be avoided by using the maximum frequency possible for a given skip distance.

A further type of fading, called **selective fading**, occurs when different parts of the signal being received take different routes through the ionosphere. This results in only part of the signal being received.

Diversity reception

This is a method used to eliminate fading (Figure 2.11). It is of two types. **Aerial diversity** involves two receiving aerials and their associated receivers spaced well apart and feeding into a common device, which accepts the strongest signal. **Frequency diversity** involves two front ends of receivers tuned to two frequencies of a common signal (a station transmitting on two frequencies – usually different bands). A selective device will accept the strongest signal from each of the front-end receivers.

Space wave or direct wave

As the name implies, this is propagation that suffers very little attenuation from the medium of the earth's surface or the ionosphere. It occurs at frequencies above 30 MHz, in the VHF range and above.

Direct wave: point to point

Space wave propagation gives ideal point-to-point communications when the receiving aerial is in sight of the transmitting aerial. This is usually referred to as **line-of-sight communications**. At these frequencies the wavelength and therefore the wavefront is small compared with objects that it may strike: if it strikes an object that is larger than its own wavelength size it will reflect off that object.

This can cause problems in the communications process, because there may be two paths that a space wave can take: direct, and indirect from a reflecting

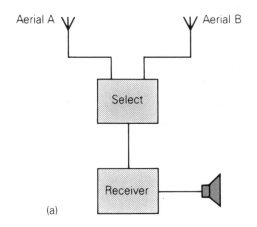

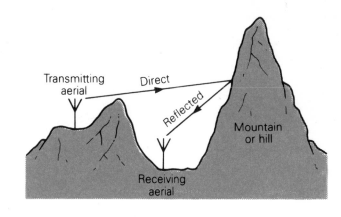

Figure 2.13 Reception via reflection.

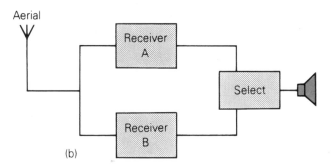

Figure 2.11 Diversity reception: (a) aerial diversity; (b) frequency diversity.

installations can suffer from weather conditions such as fog or wet weather, when attenuation becomes a problem.

There is a tendency in the lower VHF range for the space wave to refract slightly via the earth's atmosphere. This is usually compensated for by assuming that the earth's diameter is four-thirds of its natural size when considering line-of-sight communications. This makes the earth appear to have a larger surface, and therefore the line of sight will be greater.

Satellite communication uses the space wave at frequencies well above VHF and UHF, in the SHF (super high frequency) or microwave range. The units of frequency are gigahertz (GHz). There are two main reasons for using such high frequencies: (a) the number of channels that can be fitted into the spectrum available at these frequencies; and (b) the engineering and physical aspects of the size of directional aerial required at these frequencies, which is directly related to the wavelength.

Communications satellites are normally in a **geostationary orbit**. This means that they travel at exactly the same speed as the earth's rotation, and therefore appear stationary to a point on the earth's surface. Ground stations can therefore communicate with them using the direct or space wave and a directional disc aerial. If other ground stations are available to communicate, contact can be made via the satellite link.

object. These two signals may cancel each other out if they are out of phase (Figure 2.12). It can also be used to advantage when communications may be difficult because a receiving station is in a valley; a nearby mountain or hill may be able to reflect the signal into the valley (Figure 2.13).

In practice, communications systems using the point-to-point method have path profiles constructed before installation, to ensure that the paths used avoid hills and reflecting objects. In general, point-to-point communications are very reliable, although low-power

Scatter propagation

Another form of space wave communications at UHF and microwave frequencies uses disturbances in the ionosphere to reflect the wave back to earth. Two types of scatter are possible, **tropospheric** and **stratospheric**, each relying on that part of the ionosphere. The troposphere is a layer of the atmosphere extending from about 6 miles (10 km) above the earth's surface, in which temperature decreases with increase in height. The stratosphere is a layer above the troposphere, in which temperature remains constant.

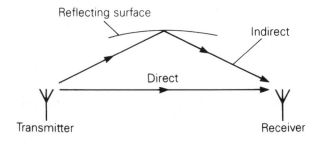

Figure 2.12 Direct and indirect paths.

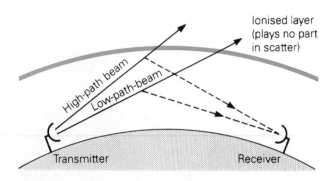

Figure 2.14 Scatter propagation.

As the amount of the wave that is returned from the scatter is small, the transmitter beam for scatter propagation to take place must be very powerful, while the receiving system must respond to weak signals (Figure 2.14). It can be considered that the high-power transmitted beam causes some disturbance in the ionosphere itself when initially set up, with some of the beam being reflected back to earth in the form of scatter, from the disturbance that has been caused.

■ CHECK YOUR UNDERSTANDING

● EM or radio propagation is possible at frequencies of 10 kHz and above.

● Aerial length is generally related to the wavelength of the EM wave.

● An EM wave induces e.m.f.s into wires that it passes or strikes.

● The signal strength of an EM wave varies in proportion to $1/d^2$, where d is the distance from the transmitter to the receiver aerial.

● There are a number of different modes of EM propagation: surface or ground wave, which stays close to the earth's surface: sky wave propagation using the ionosphere above the earth's surface to refract the EM wave back to earth, giving the possibility of worldwide communication; and space or direct wave propagation used for point-to-point, scatter and satellite communication.

● The ionosphere is a region of ionised gases at different heights above the earth. It is caused by radiation from the sun. The amount and the bands of ionised layers vary with the time of day, the seasons of the year, and sunspot activity.

● Sky wave propagation suffers from various forms of fading.

● Diversity reception can help to overcome the problem of fading.

1 Suggest reasons why it is possible to receive more radio stations on MF bands during night-time hours of reception.

2 Why is it necessary to change frequencies during 24 hours when wishing to maintain constant communication with a distant radio station via sky wave propagation?

3 Why would you expect to be able to receive a signal on MF over a greater distance across desert or seawater than over dense foliage?

4 Why are directional aerials necessary?

5 How does diversity reception help with fading problems?

6 Why is it necessary for reception purposes when receiving signals to know the polarisation of the EM wave?

7 Using a short-wave radio, tune into one of the main worldwide broadcast stations. Note the times when fading becomes excessive. Note that for good reception the frequency on which the station is received has to be changed: down as nightfall approaches and up as dawn and midday approach. Maximum frequencies for reception should occur some time after midday, while minimum frequencies should occur some time after midnight.

8 Using the same short-wave radio, try connecting two aerials that are spaced at a distance from one another, and note the results, especially during heavy fading. Does one at any one time give better reception?

9 You can demonstrate the effect of reflection and refraction of radio waves by observing the reception of these waves at different frequencies. Find a structure (a wall or a building) that is many times larger than the wavelength of a VHF broadcast radio wave that can be received using a portable receiver. If you now walk around the structure listening to the broadcast, there will be places where reception is lost. This shows that the radio wave has been reflected off the structure. At the same time you may be able to see the effect of the radio wave being reflected off a nearby structure, giving reception at that point but not at others.

You should now follow the same procedure using a portable receiver tuned to a lower frequency, perhaps at MF for a broadcast station. When you walk round the structure this time you should be able to receive the station at all times. This illustrates the effect of the radio wave refracting around the structure: the radio wavefront has a larger wavelength than the structure.

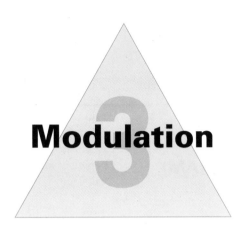

Modulation

Introduction

Spoken communication is very limited: the people communicating have to be within hearing distance of each other. The process of modulation provides a way for people to communicate over vastly greater distances. It consists in putting information – in some form or other – onto a wave of a much higher frequency. This wave then carries the information using propagation in free space, or perhaps a cable network.

Modulation can be likened to a post office system. Information is written down in the form of a letter, which is then put into an envelope (the modulation process). The envelope is addressed and sent by post (the carrier). On reaching its destination the envelope is opened (demodulation), and the contents of the letter are read. As in the modulation process, there can be more than one letter using the same route, each conveying different pieces of information.

There are several different types of modulation that can be used to convey information. They differ in complexity, and each one has its own advantages and disadvantages. We shall discuss these later in the chapter.

Bandwidth

The carrier used in the process of modulation requires space to be reserved for the information that is being conveyed – just as the bulk of an envelope will depend on the number of pages that have been written. The space on the carrier used for the information is called the **bandwidth**. It is a direct function of the amount of information to be conveyed in a particular time as well as the type of modulation used.

To illustrate this point consider an old system that is still in use for conveying messages, called **Morse code**. In this

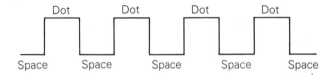

Figure 3.1 The letter H in Morse code.

system each letter of the alphabet is given a code, which is a series of dots and dashes. For example, A is dot dash, and B is dash dot dot dot. To convey a message all that is required is to interrupt a continuous carrier wave with a message using this code.

The most complicated letter in Morse code is probably H, which is four dots. To send the letter H, we have to interrupting a continuous wave eight times in the form of square waves (Figure 3.1). If we send the letter H in 1 second, this is four dots in 1 second. As a square wave may be made up of, say, 11 harmonics of the original sine wave, we shall have approximately 11×4 or $44\,Hz$ of bandwidth that the letter H has taken up. If we now increase the transmission speed by sending the dots in one fifth of a second (or send five Hs in 1 second) the bandwidth becomes 5×44 or $220\,Hz$.

This is a very important concept. The bandwidth of a signal is directly related to the amount of information required in a given time:

Bandwidth is proportional to information sent.

The more information we want to send in a given time, the more bandwidth we shall need. Morse code transmission, which sends one letter at a time (limited communication), requires little bandwidth; television, which transmits pictures and sound, requires a lot of bandwidth.

Bandwidth is important as it is this, among other issues, that limits the use of the frequency spectrum for communications.

Amplitude modulation (AM)

In this type of modulation the amplitude of the carrier wave varies in sympathy with the modulating frequency, producing an 'envelope' around the carrier frequency, which varies with the required information.

Consider a carrier frequency being amplitude modulated by a pure sine wave (Figure 3.2). At point a on the audio sine wave there will be no change in the carrier frequency amplitude. However, at point b on the audio sine wave the voltage is maximum, which causes the carrier frequency amplitude to increase to its maximum value. At point c the audio sine wave has returned to datum (0) and the carrier frequency returns to its datum value. At point d the audio sine wave is at its maximum negative value, and the carrier is reduced to its lowest value.

Percentage modulation

The amount by which the carrier frequency amplitude is varied by the modulating signal is expressed as a percentage, called the **percentage modulation**. In

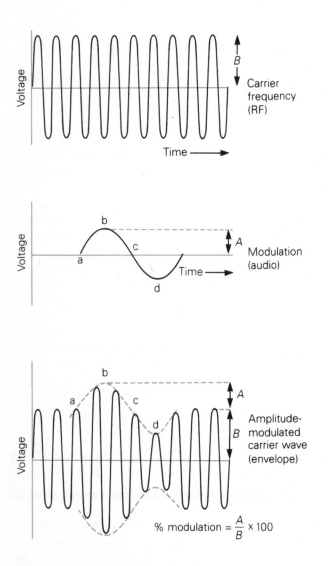

Figure 3.2 Amplitude modulation.

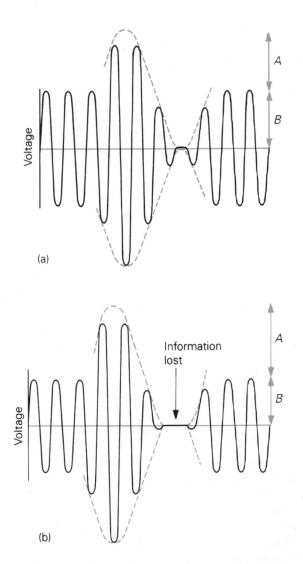

Figure 3.3 (a) If $A/B = 1$, then $1 \times 100 = 100$ per cent modulation. (b) If $A/B > 1$, then the modulation is greater than 100 per cent, and distortion occurs.

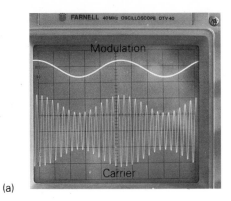
(a)

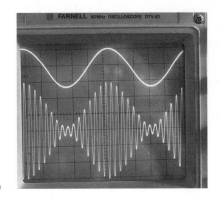
(b)

Figure 3.4 Amplitude modulation shown on an oscilloscope

Figure 3.2 the ratio of the voltage values A and B gives an index, which if multiplied by 100 gives a percentage:

$$\text{Percentage modulation} = \frac{A}{B} \times 100\%$$

The envelope of the carrier, or the positive-going excursions of the carrier, represents the intelligence or modulating signal. However, if we now increase the ratio A/B to greater than 1 – that is, make the value of the modulating signal greater than that of the carrier – some of the intelligence will be lost. The envelope of the carrier does not now faithfully represent the modulating signal. Thus the maximum percentage modulation that is possible without distortion is 100 per cent (Figures 3.3 and 3.4).

Sidebands

It can be shown mathematically that when a carrier wave (f_c) is amplitude modulated by another signal (f_m), three separate frequencies are produced:

1. the carrier frequency, f_c;
2. an upper side frequency or range of frequencies called the **upper sideband**, $f_c + f_m$;
3. a lower side frequency or range of frequencies called the **lower sideband**, $f_c - f_m$.

Consider a carrier with a frequency of 1 MHz, amplitude modulated by a frequency of 1000 Hz (Figure 3.5a). This would produce three separate frequencies: 999 000 Hz (the lower side frequency), 1 000 000 Hz (the carrier), and 1 001 000 Hz (the upper side frequency).

In practice, the modulating frequency is usually a range of frequencies. For audio frequencies this may be in the range 300–3400 Hz. Consider now this range of frequencies modulating a 4 MHz carrier (Figure 3.5b). This would produce a carrier frequency of 4 MHz, an

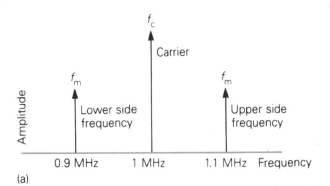
(a)

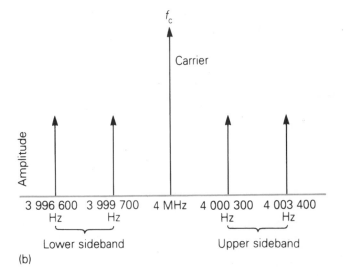
(b)

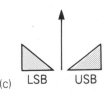
(c)

Figure 3.5 (a) Side frequencies; (b) sidebands. (Note that the x axes are not to scale.) Upper and lower sidebands are sometimes shown as in (c).

upper sideband of frequency 4 MHz + (300–3400 Hz), and a lower sideband of frequency 4 MHz − (300–3400 Hz).

Bandwidth

From the above we can see that we can obtain the bandwidth of an AM wave by subtracting one sideband (or side frequency) from the other. That is, $(f_c - f_m)$ minus $(f_c + f_m)$ gives $2f_m$. Therefore the bandwidth of an AM wave is twice the highest frequency of the modulating waveform.

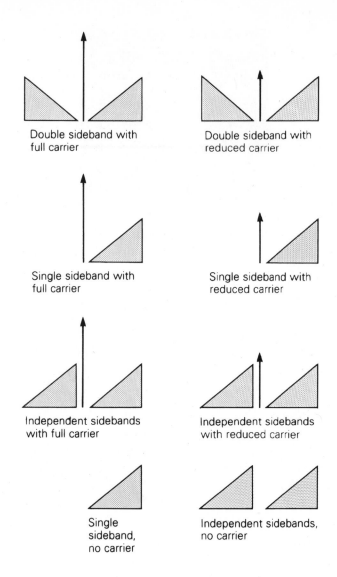

Figure 3.6 Different types of AM transmission.

The figure shows: Double sideband with full carrier; Double sideband with reduced carrier; Single sideband with full carrier; Single sideband with reduced carrier; Independent sidebands with full carrier; Independent sidebands with reduced carrier; Single sideband, no carrier; Independent sidebands, no carrier.

EXAMPLE

An 8 MHz carrier wave is amplitude modulated by a band of audio frequencies 300–3400 Hz.

Determine: (a) the frequencies contained in the modulated wave, and (b) the bandwidth occupied by the signal.

(a) f_c = 8 MHz
   LSB = 8 MHz − (300 to 3400 Hz) = 7 996 600 to 7 999 700 Hz
   USB = 8 MHz + (300 to 3400 Hz) = 8 000 300 to 8 003 400 Hz.
(b) Twice the highest frequency of modulation = 2 × 3400 = 6800 Hz.

Single-sideband AM transmissions

The carrier frequency component of the AM waveform produces no information, while the two sidebands that are produced in amplitude modulation, when processed by demodulation (removing unwanted frequencies from the AM wave) produce the same information. There is thus a case for using only the component containing the information: that is, the upper or lower sideband. This has several advantages:

1. The transmitted signal takes up less room for a given amount of information: that is, it requires less bandwidth.
2. The transmitted signal requires less power to be produced; alternatively, all the power can be directed into the single-sideband (SSB) signal.
3. The carrier frequency can be dispensed with, or its amplitude can be reduced, thus reducing the power required in the transmitter output stages.
4. Two different sidebands can be transmitted together, each containing different information. Figure 3.6 illustrates different types of AM transmission.

Mathematical approach to amplitude modulation

The general expression for a sinusoidal carrier wave is

$$v = V_c \sin(\omega_c t + \phi)$$

where v is the instantaneous carrier voltage, V_c is the peak or amplitude of the carrier, ω_c is $2\pi \times$ carrier frequency, ϕ is the phase of the carrier voltage at time zero, and t is the time in seconds.

When a carrier wave is amplitude modulated the amplitude of the carrier voltage will be varied in accordance with the characteristics of the modulating signal:

$$v = V_m \sin \omega_m t$$

where v is the instantaneous value of the modulating signal, V_m is the peak value of the modulating signal, ω_m is $2\pi \times$ frequency of the modulating sine wave, and t is the time in seconds. This indicates that the mean value of the AM wave will be V_c, while the peak value of the variation from this mean value V_c will be V_m. The frequency of this variation will be ω_m.

Thus the instantaneous value of an amplitude-modulated wave is given by

$$v = (V_c + V_m \sin \omega_m t) \sin \omega_c t$$

This is an important general expression for an AM wave.

If we now multiply this out:

$$v = V_c \sin \omega_c t + V_m \sin \omega_m t \sin \omega_c t$$

Now using the compound angle trigonometric identity

$$2 \sin A \sin B = \cos(A - B) - \cos(A + B)$$

where $A = \omega_c t$ and $B = \omega_m t$, we get

$$v = \underbrace{V_c \sin \omega_c t}_{\text{Carrier}} + \underbrace{\frac{V_m}{2}\left[\cos(\omega_c t - \omega_m t)\right]}_{\text{LSF/LSB}}$$

$$\underbrace{- \frac{V_m}{2}\left[\cos(\omega_c t + \omega_m t)\right]}_{\text{USF/USB}}$$

This indicates that there are three separate frequency components of an amplitude-modulated wave:

1. the carrier wave;
2. a lower side frequency or sideband;
3. an upper side frequency or sideband.

Maximum amplitude occurs when $\sin \omega_m t = 1$, i.e. when $V_c + V_m = 2$.

Minimum amplitude occurs when $\sin \omega_m t = -1$, i.e. when $V_c + V_m = 0$.

This condition represents 100 per cent modulation, or a modulation index of 1.

Disadvantage of AM

In the condition above – the maximum modulation that can be tolerated without distortion – it can be mathematically proved that the power in the sidebands represents only half the power in the carrier part of the composite signal. In other words, only a quarter of the power transmitted from a 100 per cent modulated waveform is in each of the sidebands. This represents a considerable waste of power, as the useful information to be transmitted is contained only in one sideband.

Reception of AM also suffers from amplitude variations such as **impulse noise** (spikes of voltage caused by sparks when electricity is switched on and off), which would be detected by an AM receiver and be presented as an output of noise in the loudspeaker.

Angle modulation

This term is generally used to describe two types of modulation, referred to as **frequency modulation** and **phase modulation**, which are used to overcome some of the disadvantages of amplitude modulation. As both use carriers of constant amplitude, and both occur if one is enacted, they are usually discussed together.

Frequency modulation

Frequency modulation uses a carrier of constant amplitude, whose instantaneous frequency is varied in sympathy with the modulating frequency. Figure 3.7 shows the principles of frequency modulation. The amount by which the carrier frequency is varied either side of its 'norm' frequency is called the **deviation** or **frequency swing** (Δf), and this is determined by the amplitude of the modulation or loudness (volume) of the modulating frequency.

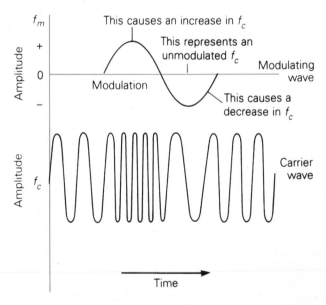

Figure 3.7 Frequency modulation.

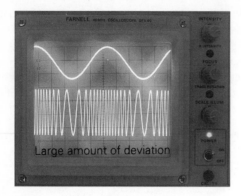
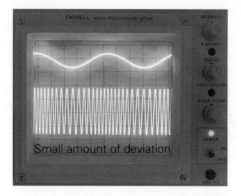

Figure 3.8 Frequency modulation: the extent of deviation represents the amplitude of the modulation.

The rate of change of the carrier frequency is determined by the frequency of modulation. If the modulating frequency is low, the rate of change will be small; if the modulating frequency is high, the rate of change will be high. Therefore the rate of deviation of the carrier represents the frequency of the modulation, while the extent of the deviation represents the amplitude of the modulation (Figure 3.8).

The **modulation index** of a frequency-modulated waveform is defined as follows:

Modulation index, m_f

$$= \frac{\text{Deviation of carrier frequency, } \Delta f}{\text{modulation frequency, } f_m}$$

The bandwidth of an FM waveform is more difficult to determine than that of an AM waveform because, in this type of modulation, the angular bandwidth consists of a number of sidebands $\omega_c \pm n\omega_m$, where n is an integer (whole number). This involves complicated mathematical analysis, which is outside the scope of this book. Suffice it to say that the bandwidth occupied by an FM waveform is considerably larger than that occupied by an AM waveform. We can obtain an indication of the bandwidth by using the following formula:

$$f_{bw} = 2(\Delta f + f_{m(max)})$$

where Δf is the frequency of deviation, and $f_{m(max)}$ is the maximum modulating frequency.

EXAMPLE

Calculate the approximate bandwidth of a broadcast transmission with a deviation of 75 kHz and a maximum modulation frequency of 12 kHz. Compare this bandwidth with that of a communications VHF channel using a deviation of 25 kHz and a maximum modulation frequency of 3.4 kHz.

Using the formula above, the broadcast bandwidth works out at 174 kHz, while the communication channel bandwidth works out at 56.8 kHz.

In practice, frequency modulation operates with a maximum deviation that corresponds to 100 per cent modulation in an AM waveform. This is due to the bandwidth requirements for FM. For broadcasts this maximum deviation may be in the order of ± 75 kHz; for communications, where the quality of the broadcast is not so important, this can drop to ± 25 kHz, thereby providing more space in the frequency spectrum for other communication channels. Hence the use of FM only on VHF or above, where there is more space in the frequency spectrum.

Phase modulation

This type of modulation alters the instantaneous phase relationship of the carrier to its 'norm' in sympathy with the modulating frequency. In this respect it is easy to see the close relationship between frequency modulation and phase modulation, because by changing one you must also change the other. Therefore the characteristics of frequency modulation must apply to phase modulation; most demodulators in FM receivers are conveniently able to work with both types of modulation.

An adaptation of this type of modulation, used in modern digital communication and TV circuitry, is referred to as **phase shift keying**. There are two types: binary phase shift keying (BPSK) and quadrature phase shift keying (QPSH). In BPSK the modulating waveform has two states, +1 and –1, which correspond to the phase of the carrier waveform being advanced or retarded by +90° or −90° ($\pi/2$ radians). This gives a two-state modulation stream, which is ideal for digital transmissions. In the more complicated QPSK, the phase of the carrier is set by the modulation to one of four possible values: $\pi/4$, $3\pi/4$, $5\pi/4$ and $7\pi/4$.

Advantages of angle modulation

Noise in electronics usually produces amplitude variations of the carrier, normally in the form of 'spikes'. It is therefore advantageous to use a system that relies on a constant-amplitude signal. Indeed, in FM and phase receivers, limiters are included in the circuitry to ensure that the carrier wave produces a constant-amplitude signal to the demodulator circuit. This means that FM and phase modulation suffer less interference from noise than the equivalent amplitude modulation.

Another advantage associated with FM reception is referred to as **capture effect**. This is noticeable when tuning-in FM stations on receivers. A station is either 'captured' or not: there is no in-between stage, as there may be with AM reception. Either noise is present or the signal is present. This means that there is either very good reception of a station or no reception at all.

Disadvantages of angle modulation

The main disadvantage is the amount of bandwidth that this type of modulation uses. It is not normal practice to use FM or PM on the lower-frequency bands: the use of angle modulation is restricted to VHF and above. This means that the signals will be limited in range for terrestrial transmission (see Chapter 2 on propagation). This disadvantage does not apply to satellite communication of course.

Transmitters and receivers both tend to have more complex circuitry than for AM. This is possibly outweighed by the fact that the power amplifier circuits, which work at constant amplitude, can be made more efficient.

Pulse modulation

As the name suggests, this type of modulation uses pulses to transmit information. These can be either pulse amplitudes or pulse widths. Whichever system is used, some form of sampling must be applied to the original information; this is then transformed into pulses that correspond to the original information.

As can be seen in Figure 3.9, the process involves a sampling stage. If the interval between samples is too long then some of the information being sampled will be lost, as the resultant decoding in the receiver will not have enough information to reproduce the original fully. Theory tells us that the signal may be reconstructed without error from regularly spaced samples taken at a rate that is at least twice the highest frequency being sampled. The transmitting of samples may take more bandwidth than the original signal, but the space between samples can be used for other samples from other signals. This system of space and time sharing is called **time division multiplexing** (TDM), and is outside the scope of this book.

Because the pulses produced in this type of modulation can be subject to amplitude distortion during transmission, the system is further refined by using a system called **pulse-code modulation** (PCM). In this system the results of the sampling are **quantised** – resolved into one of a finite number of possible values.

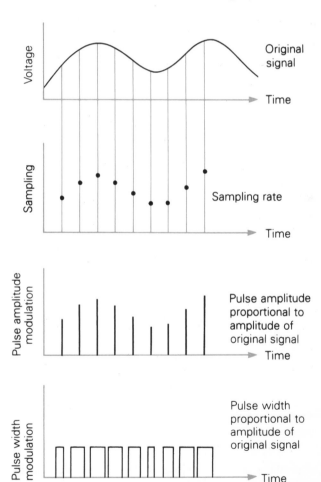

Figure 3.9 Pulse amplitude modulation (PAM) and pulse width modulation (PWM).

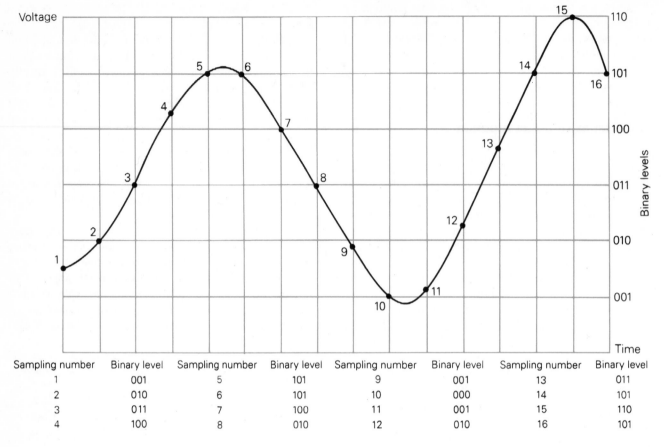

Sampling number	Binary level	Sampling number	Binary level	Sampling number	Binary level	Sampling number	Binary level
1	001	5	101	9	001	13	011
2	010	6	101	10	000	14	101
3	011	7	100	11	001	15	110
4	100	8	010	12	010	16	101

Figure 3.10 Quantising and binary levels.

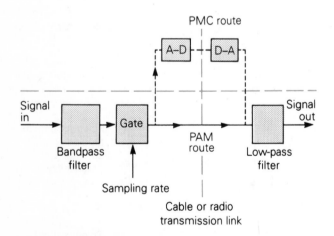

Figure 3.11 A typical link using PAM and PCM techniques.

The quantised value chosen is that represented by the band that the sampled analogue signal has fallen into (Figure 3.10).

These quantised values are then binarily coded as separate digital words using an analogue-to-digital converter, and transmitted as binary codes. The effectiveness of the process is dependent on the correct number of levels and the sampling rate. At the receiver end the process is reversed, with a digital-to-analogue converter. A measure of the effectiveness of the process introduced by the quantising process is termed **quantisation error**, and the reconstructed analogue signal experiences added noise referred to as **quantisation noise**. Figure 3.11 shows a typical link using PAM and PCM techniques.

■ CHECK YOUR UNDERSTANDING

● Modulation is the process of putting information onto a wave of another frequency.
● Bandwidth is the amount of frequency space used in the frequency spectrum when a wave has been modulated.
● Bandwidth is directly proportional to the information carried by the wave in a given time.
● Radio signals can be modulated in a number of different ways. Two popular methods are amplitude modulation (AM) and frequency modulation (FM).

- AM is formed by varying the amplitude of the carrier wave.
- FM is formed by varying the frequency of the carrier from its norm.
- AM requires high-power output stages, while FM requires generally more complex circuitry.
- FM requires a larger bandwidth than the equivalent AM signal.
- Phase modulation is similar to FM and can generally use the same demodulation circuits.
- Pulse modulation is used for digital communications.
- Pulse-coded modulation (PCM) transmits binary (digital) words to represent sampled parts of the modulating waveform.

REVISION EXERCISES AND QUESTIONS

1 Explain what is meant by the term 'modulation'.
2 Use diagrams to show the effect of AM on a carrier wave when modulated to (a) 50 per cent and (b) 100 per cent. Explain why AM of speech should not exceed 100 per cent.
3 Give the three separate components produced when a carrier wave is amplitude modulated. If a carrier frequency is 16 MHz, and it is modulated by a range of frequencies from 350 to 4000 Hz, give the components of the resultant waveform.
4 Draw a sketch to illustrate the process of frequency modulating a carrier wave.
5 Discuss the advantages and disadvantages of AM and FM.
6 Explain the term 'sampling' as applied to pulse amplitude modulation.
7 In what form is PCM usually transmitted?
8 Why is FM normally only used with frequencies at or above VHF?

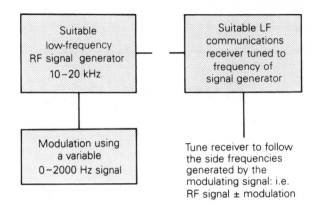

Figure 3.12 Equipment for exercise 10.

9 Using an FM radio with the squelch control switched off, confirm the presence of the capture effect by tuning through the frequency of a strong FM broadcast station. Note the absence of any intermediate stage between hearing the station and just hearing noise. Now try tuning through a weaker station: you should notice the same effect.

Now go through the same procedure using an AM radio and an AM broadcast station. Note the different effect: the presence of signal *and* noise during tuning indicates that capture effect does not occur when using AM.

10 You can readily observe the effects of amplitude modulation and bandwidth using the equipment shown in Figure 3.12 together with a communications receiver that can receive the low-frequency RF signal. A suitable RF signal generator could be an Airmec type, which has a mechanical sweep control for the modulation. Using this to sweep the modulating signal slowly between 0 and 2000 Hz should enable you to move the tuning of the communications receiver to follow the sideband either side of the carrier.

Basic transmission techniques

Introduction

In any communications system or network, whether it be simple point-to-point communication or a television system, there has to be a transmission point and a corresponding reception point or points. The reception end of the communications network is called a **receiver**. One transmission is received by one or many receivers. The device at the transmission end is called a **transmitter**. One transmitter can serve one or many receivers. This chapter looks at some basic transmission techniques, and the reasons behind the different designs of transmitter.

Transmitters are designed for specific purposes. They can be small, such as those built into modern mobile phones, which serve as both transmitter and receiver (transceiver), or large, such as those used as broadcast transmitters, where the output stage may have to be housed in a separate building and be water cooled. They do, however, have some essential features, and these will be investigated in this chapter.

The basic transmitter

A basic continuous-wave (CW) transmitter requires a device to produce a sine wave at the required frequency of transmission, which can then be switched on and off at strategic times to produce some intelligence (this could be the Morse code). Figure 4.1 shows a simple CW transmitter.

At this stage it is useful to consider what device or devices can be used to produce a suitable radio frequency that can be used for transmission. One such useful device, called a **parallel-tuned circuit**, consists of an inductor and a capacitor (Figure 4.2). This circuit has two very useful properties used in RF transmitter and receiver design:

1. The components used can store and release energy, transferring this energy between themselves at a frequency that depends on the values of the inductance and capacitance used.
2. The circuit has the ability to offer maximum opposition to an RF signal at one frequency or a small band of frequencies.

The first property mentioned is the one that we can use to produce an RF signal. If the circuit consisted of a pure inductance and capacitance, then once it was excited (or

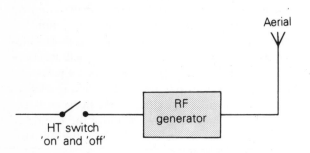

Figure 4.1 A simple continuous-wave (CW) transmitter.

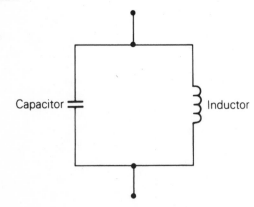

Figure 4.2 The parallel-tuned circuit.

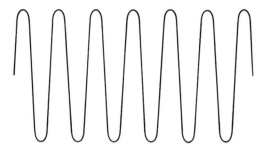

Figure 4.3 Output across either component of the parallel-tuned circuit, assuming that purely reactive components are used.

energy was injected into it), the two pure reactances would store and release energy in turn, thereby creating a stable sine wave output across either of the components (Figure 4.3). The circuit would then be said to be **oscillate**.

The frequency of this oscillation can be calculated by the following formula:

$$\text{Frequency of oscillation} = \frac{1}{2\pi\sqrt{LC}}$$

where L is the inductance in henrys, and C is the capacitance in farads.

Unfortunately, inductors and capacitors cannot be made purely reactive, and are always subject to some losses. These, together with the resistance of the connecting medium, and the fact that connecting other circuits across the components will introduce losses, can all be represented by resistance in the parallel circuit (indicating losses). Our free-running oscillation is now converted to a **damped oscillation**, with the frequency being slightly altered by the inclusion of the losses.

$$\text{New frequency of oscillation} = \frac{1}{2\pi\sqrt{\left(\dfrac{1}{LC} - \dfrac{R^2}{4L^2}\right)}}$$

where L is the inductance, C is the capacitance, and R is the resistance and losses (Figure 4.4).

To replace the losses in the circuit, and to prevent the oscillation dying out into a damped oscillation, an active device is added to 'top up' the energy lost in the circuit. This also compensates any losses that may occur when the circuit is used for transmission. This new circuit, consisting of the RF-controlling circuit (the parallel-tuned circuit in this case), the amplifier (active device) and an output terminal, is called an **oscillator** (Figure 4.5).

A transmitter for satisfactory communications needs to have an oscillator that can produce a sine wave at a constant frequency so that reception can be obtained reasonably easily. It would be inappropriate if the

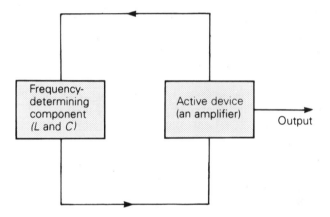

Figure 4.5 Schematic diagram of an oscillator.

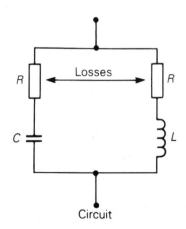

Circuit

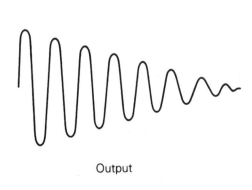

Output

Figure 4.4 Actual parallel-tuned circuit with damped oscillation output.

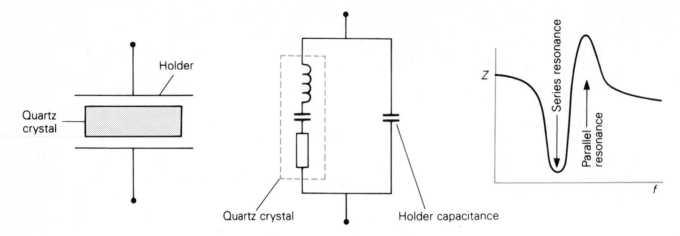

Figure 4.6 The equivalent circuit of a quartz crystal.

receiver had to be constantly retuned in an effort to follow an unstable transmission frequency.

The ability of the oscillator to remain stable on the required frequency is referred to as its **stability**. This will vary if the value of the inductance, capacitance or losses (resistance) changes. Thus the design of a stable oscillator must involve keeping these three components at a constant value. We can replace the parallel-tuned circuit with a quartz crystal. This eliminates the possible variations in inductance, capacitance and resistance. The result is called a **crystal-controlled oscillator**.

When a voltage is applied across the face of a quartz crystal, it tends to distort or twist along its axis. Conversely, when it is twisted or distorted, an electric

voltage appears across its face. This phenomenon is called the **piezoelectric effect**. The equivalent circuit of a quartz crystal looks very much like the parallel-tuned circuit in its parallel resonance mode. The inductance and capacitance are 'built in', and not subject to change: so the crystal can be applied in an oscillator with improved stability.

From the response curve of the quartz crystal shown in Figure 4.6 we can see that there are two modes in which it can be used: a **series mode** and a **parallel mode**. We have already discussed the latter mode, and Figure 4.7 shows a possible crystal-controlled oscillator circuit. In this example the crystal is acting in the parallel-tuned circuit mode. The tuned circuit in the output is tuned slightly

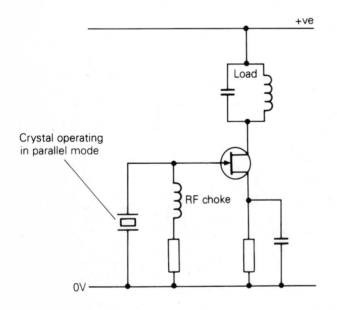

Figure 4.7 Miller crystal-controlled oscillator.

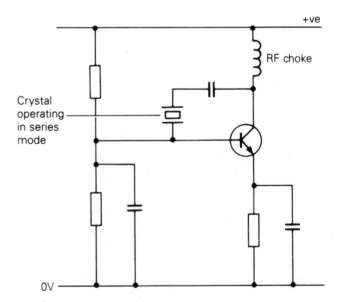

Figure 4.8 Pierce crystal oscillator.

above the crystal frequency. Feedback of energy takes place via the internal capacitance between the FET drain and gate, C_{gd}, which has been increased by the so-called **Miller effect**. This effectively increases C_{gd} by $(1 + A)$, where A is the gain of the stage.

The series mode of operation can also be employed as a direct feedback path between output and input (Figure 4.8): this oscillator is generally referred to as a **Pierce crystal oscillator**. In this circuit only one frequency can be fed back to the input circuit: the series resonance frequency of the crystal. This is called an **accepter circuit**. The circuit becomes regenerative, sustaining oscillations at that frequency.

The frequency of an oscillator can also be affected by the temperature of the oscillator's components as well as the value of the power supply and the load that the oscillator works into. Thus we would expect to find the main or master oscillator of a transmitter to be temperature controlled, to have a stable power supply, and to work into a constant load.

A typical CW transmitter

Figure 4.9 shows the block diagram of a typical CW transmitter.

Master oscillator

This is a crystal-controlled oscillator. It is normally kept at a constant temperature, and in class A mode to ensure frequency stability. To change frequency, another crystal is normally switched into the circuit.

Buffer amplifier

This stage provides the oscillator with a constant load. It should therefore be a stage with a high input impedance.

The load for this stage is usually a parallel-tuned circuit, so that maximum load and therefore amplification occurs when the load is tuned to the frequency of the master oscillator. Occasionally the buffer or a further stage acts as a frequency doubler, feeding the output amplifier with twice the master oscillator output frequency. Further stages can be added to further multiply the output frequency from the master oscillator: in this way more than one transmission frequency can be obtained from the same crystal oscillator. Note that the frequency multiplication stages would operate in non-linear class C mode.

Output stage

This stage further amplifies the signal to a power suitable for transmission into the aerial. It is usually a power stage and therefore dissipates heat (the power that is not transmitted into the aerial). Modern medium-power transmitters use power FETs for this purpose. However, large power transmitters still use thermionic valves, which may involve air or water cooling (see later in this chapter).

Matching device

The output stage has to be 'matched' into the aerial to ensure that maximum power transfer can take place between the output stage and the aerial. There are a number of different circuits used for this purpose depending on the frequency of transmission and the aerial used. Matching requirements are discussed later in this chapter. Their basic purpose is to make the output stage 'think' it is operating into the optimum load, and to couple this load into a suitable aerial. Thus the circuit used is normally a compromise.

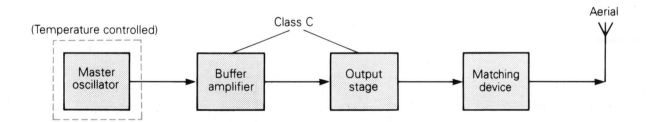

Figure 4.9 Block diagram of a typical CW transmitter.

The amplitude-modulated (AM) transmitter

In an AM transmitter, the amplitude of the RF output from the transmitter is varied in sympathy with the intelligence that is required to be carried by the output wave. The easiest way to achieve this is to vary the amplification of one of the blocks that make up the transmitter we have already discussed. This can be done by varying the power supply to the output stage in sympathy with the intelligence required, or by using a more complex mixing circuit, which can be used to generate single sideband, independent and/or reduced carrier type signals in the early stages of the transmission process.

Which method is best? There are advantages and disadvantages both in modulating the transmitter in the early stages (**low-level modulation**), and in leaving it to the final stage (**high-level modulation**).

High-level modulation

Figure 4.10 shows the block diagram of an AM transmitter using high-level modulation. The stages up to the power output stage are the same as discussed above for the CW transmitter, so we shall discuss only the additional block stages.

Microphone or other suitable intelligence
This is fed at a low level to the next stage. As such the signal produced may be of the order of microvolts rather than volts.

Preamplification
This stage amplifies the output from the low-level input and possibly processes it into a form suitable to be modulated. For instance, it may be compressed to restrict the modulation to a specific frequency band. For example, telephone conversations are restricted to the frequency band 350–3400 Hz.

AF amplifier
This stage provides the power needed to drive the power amplifiers that are required to supply high-level modulation. In this case the stage might be two class B amplifiers in push–pull configuration.

Modulator
This provides the power necessary to alter the envelope of the final stage. For 100 per cent modulation this stage must provide half the power of the waveform being transmitted. It is normally a class B stage in push–pull, working into a transformer giving the required linearity and an improved efficiency. Figure 4.11 shows the usual final stage arrangement.

The advantages of high-level modulation are simplicity and ease of circuit design: stages can work at maximum efficiency mode up to and including the final stage. The disadvantage is that modulation takes place at a high power, and therefore the modulator is required to be high power also (supplying half the power at 100 per cent), making layout design difficult and components expensive due to insulation problems.

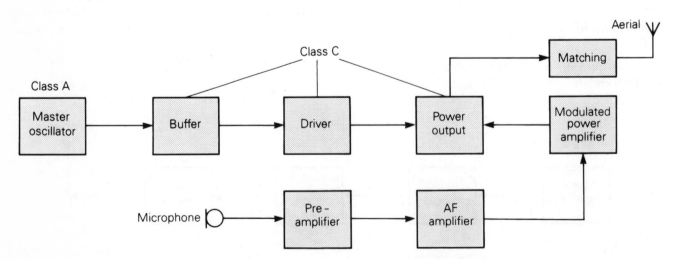

Figure 4.10 Block diagram of an AM transmitter using high-level modulation.

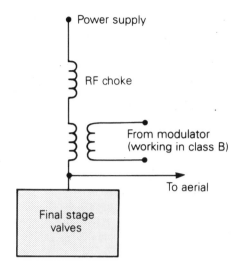

Figure 4.11 Final stage arrangement for high-level modulation.

Low-level modulation

Figure 4.12 shows the block diagram of an AM transmitter using low-level modulation. Modulation now takes place at an early stage in the transmission process, and therefore does not require so much power as in high-level modulation. However, the stages following the modulator have to work in a linear mode and therefore are not as efficient, creating a problem with heat dissipation.

There are a number of different designs for the modulator or mixer circuit. The one shown in Figure 4.13 is called a **balanced modulator**. The circuit normally includes balancing components to compensate for capacitance and inductance in the circuit. The carrier signal is arranged to be much greater than the modulating signal. The carrier signal switches the diodes in such a way that it is cancelled out. The signal

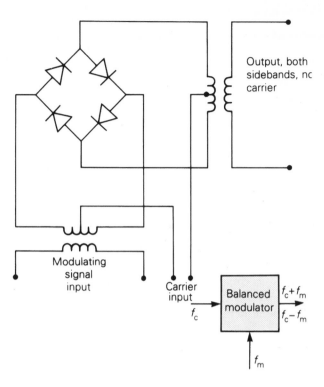

Figure 4.13 Balanced modulator: basic ring or lattice type modulator.

output is then the carrier frequency plus and minus the modulating signal.

For double-sideband or reduced-carrier transmission the carrier has to be re-injected into the circuit at a later stage. However, for single-sideband transmission the unwanted sideband can be eliminated using a filter circuit before passing to the next stage. Similarly a further sideband can be added containing different information to replace the unwanted sideband, creating independent sideband transmission.

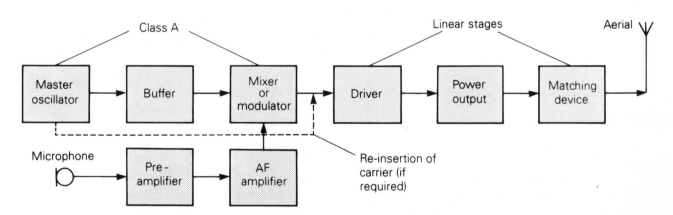

Figure 4.12 Block diagram of an AM transmitter using low-level modulation.

The advantages of low-level modulation are that all modulation is done at low power, making circuit layouts smaller and easier to design, and that single sideband and other forms of modulation are easily made available. The disadvantages are that the circuits become increasingly complex, and that the stages following modulation are required to be linear, thereby creating a problem with heat due to lack of efficiency in the final stages. Large transmitters would therefore have to use thermionic valves in their final stages, possibly with air blast or water cooling.

The frequency-modulated (FM) transmitter

The basic circuit follows that of the transmitter types discussed above. However, in this case it is an advantage that the amplitude of the waveform being transmitted through the transmission blocks remains constant. This means that the modulation is done at an early stage, very commonly in the carrier oscillator itself by means of varying the instantaneous value of the oscillator output in sympathy with the modulating signal. It is normal to have an automatic frequency control (AFC) loop to ensure that the centre frequency of the transmission is controlled by a crystal-controlled oscillator.

Figure 4.14 shows the block diagram of a typical FM transmitter.

AF amplifier: pre-emphasis

This stage amplifies the audio intelligence and produces a corresponding d.c. voltage, which operates the varactor in the modulator. There is a tendency in FM for the weaker, higher frequencies of the audio spectrum to be submerged into **white noise** (random electronic noise that exists across the frequency spectrum) compared with the stronger low audio frequencies. A special circuit is therefore introduced into FM audio circuits to

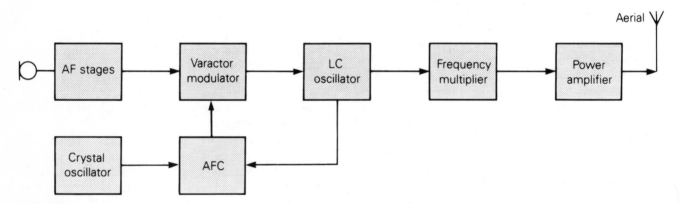

Figure 4.14 Block diagram of an FM transmitter.

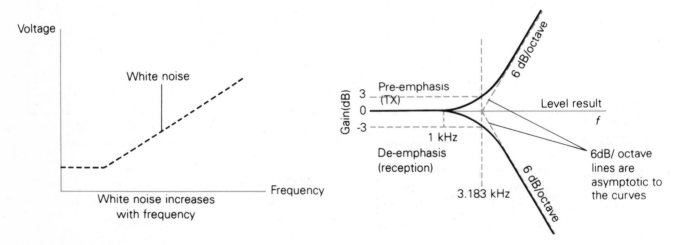

Figure 4.15 Pre-emphasis and de-emphasis.

compensate for this effect. In the transmitter this is referred to as **pre-emphasis**.

The process consists in amplifying the high-frequency part of the audio spectrum more than the lower frequencies. This is normally accomplished using a reactive-resistor circuit giving a 6 dB/octave rise in amplification. In the receiver the reverse process takes place in the audio amplifier, with the low frequencies being amplified more than the high frequencies. This is known as **de-emphasis**. In this way the signal-to-noise ratio is improved at the higher audio frequencies by reducing the white noise levels at those frequencies. Figure 4.15 illustrates this principle.

LC oscillator

This provides the modulation process in the transmitter. Its instantaneous oscillator output is controlled by a varactor modulator. The centre frequency of the oscillator is stabilised by the use of a crystal-controlled oscillator via an automatic frequency control (AFC) loop. The amount by which the frequency deviates from the carrier is small at this stage.

Frequency multiplication stage

The output from the LC oscillator, which has its instantaneous value changed slightly by the varactor modulator, is multiplied in a series of class C stages until it reaches the transmitted frequency.

Power amplifier

The power amplifier – another class C amplifier – provides the power to be transferred into the aerial circuits.

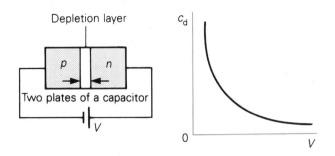

Figure 4.16 The principle of the varactor diode. The junction capacitance is proportional to $1/\sqrt{V}$.

Varactor modulator

This uses the variable capacitance of a *p-n* junction semiconductor known as a **varactor diode** (Figure 4.16). When a d.c. voltage is applied in the reverse bias direction across the *p-n* junction, the charges are drawn away from the junction, leaving a depletion layer (an area depleted of carriers – that is, an insulator). This represents a capacitance, which appears across the junction between the separate charges and may be varied by varying the applied d.c. voltage (Figure 4.17). If the varactor diode is now made part of the parallel-tuned LC circuit, the capacitance and therefore the frequency of the tuned circuit can be made to vary in sympathy with a d.c. voltage from the AF amplifier.

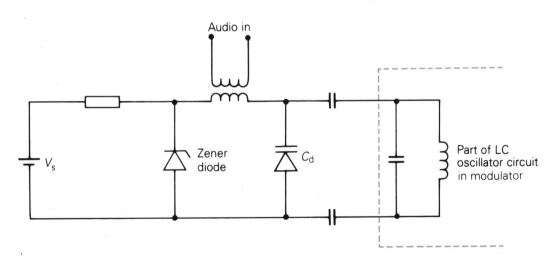

Figure 4.17 The use of the varactor diode as part of the modulator.

The need for power stages

We have already mentioned the problems associated with heat in the final stages of the transmitter, especially if the output valves are having to work in a linear mode, when an efficiency of 15–25 per cent would be considered good. This means that 75 per cent of the power in the final stages has to be dissipated in the devices themselves. Modern advances in technology have meant that special power FETs have been developed for low- and medium-power transmitters, which have special provision for dispersing heat, and special cooling fins have been devised. However, for high-power transmitters thermionic valves are still used. They employ air blast and water cooling to overcome the effects of the heat from the surplus power in these valves.

Power output valves (usually power triodes) and FETs share a common problem: they both have a capacitance inherently built into their structures between the input terminal and their output terminals if used in a common amplifier configuration. Typical values are 1–5 pF for valves and 3–7 pF for transistors. The Miller effect, mentioned earlier in connection with the Miller crystal oscillator, increases the effective capacitance value to (1 + amplification factor of the device) multiplied by the actual value of the capacitance. This can cause problems at high frequency: even the smallest of capacitance values can cause instability in the form of feedback from the output circuit to the input circuit of the device via the internal capacitance (Figure 4.18).

In most output circuits this possible instability is overcome by feeding back energy 180° out of phase via

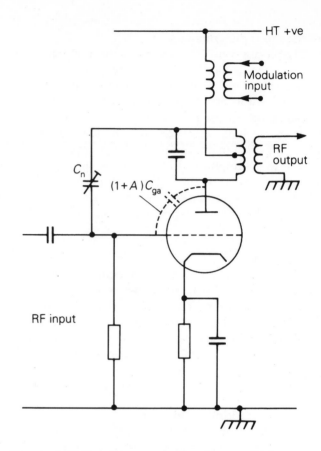

Figure 4.19 Hazeltine neutralisation applied to a power output thermionic valve.

what is referred to as a **neutralising capacitance** or **neutralising circuit**, thereby cancelling out the effect of the internal capacitance. Figure 4.19 illustrates one form of neutralisation, called **Hazeltine neutralisation**. C_n is adjusted to be the same value as $(1 + A)C_{ga}$, where C_{ga} is the internal capacitance between the grid and anode of the valve.

The internal structure of large output valves alters during their lifetime because of heat stresses. This can alter the internal capacitance, so that the neutralising capacitance may have to be varied. This results in a regular maintenance task for the engineers who service transmitters.

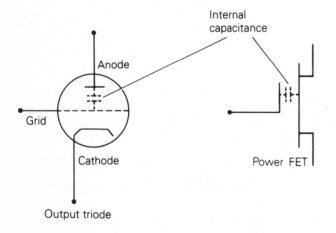

Figure 4.18 Internal capacitance in power output devices. It is effectively multiplied by $(1 + A)$, where A is the amplification factor of the device (valve or FET).

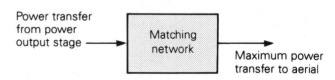

Figure 4.20 Matching requirements for the output stage.

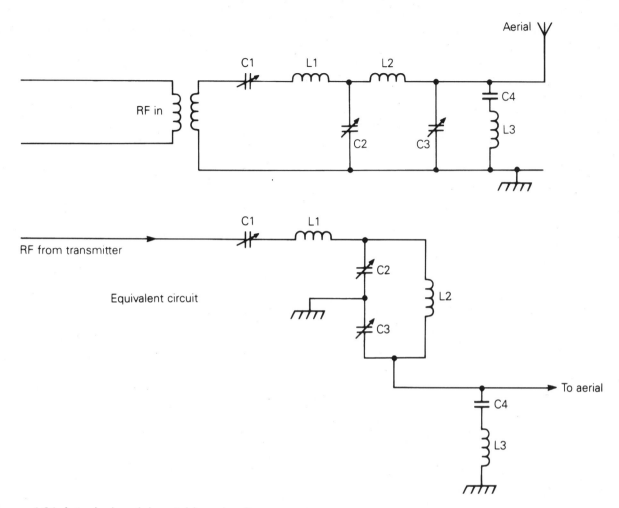

Figure 4.21 A typical aerial-matching circuit.

Output circuits: loading and matching to aerial

Figure 4.20 illustrates the matching requirements needed to feed maximum power from the output stage into the aerial. The basic impedance of the aerial will change depending on the frequency being transmitted for a given aerial length.

Figure 4.21 illustrates a typical matching circuit. As several of the components are interrelated, the procedure for obtaining maximum transfer of power into the aerial may involve repeating the 'tuning' of components a number of times until no further increase in aerial current is indicated.

In the circuit shown, C1 and L1 form a series-resonant tuned circuit (**accepter circuit**). This is tuned to the output frequency of the transmitter, and therefore offers a low-impedance path for that frequency only, effectively stopping all other frequencies.

C2 and C3 form a parallel-tuned circuit with L2. Adjusting C2 will alter the impedance seen by the output from the final stage. This should be adjusted for maximum power transfer from the output stage into this circuit – usually indicated by a dip in the current being measured flowing through the final stage. By adjusting C3, the aerial can be matched into this circuit, usually indicated by a meter measuring an indication of aerial current. Optimum results need progressive adjustments between C2 and C3. C4 and L3 form a series-tuned circuit that effectively short-circuits second-harmonic transmission frequencies to earth, and avoids transmission at this frequency from parasitic oscillations, which may occur in the output circuits.

In higher-frequency transmitters (UHF and above) there may be difficulty in recognising reactive components (that is, inductors and capacitors) because of the low values required: at these frequencies a short piece of wire can represent an inductor, or when run alongside another wire or earth may represent a capacitor. You should bear this in mind when examining the output stages of transmitters, and tuners in receivers, when using UHF and above.

■ **CHECK YOUR UNDERSTANDING**

● Parallel-tuned circuits have two useful properties when used in RF transmitter and receiver design: (a) the components store and release energy; (b) the circuit offers maximum opposition to an RF signal or small band of frequencies. It does this at the approximate frequency of $1/(2\pi\sqrt{LC})$.

● A quartz crystal can operate in two modes: as an equivalent series-tuned circuit and as a parallel-tuned circuit. The frequencies at which these modes operate are close together in the frequency spectrum.

● An oscillator's ability to remain stable on one frequency is termed its stability. Stability is improved by using crystal oscillators, by keeping the components at a constant temperature, by using a stabilised power supply, and by feeding the output into a constant load.

● Power dissipation can be a problem in the output stages of a transmitter. It normally requires special attention in the design of the final stages. Power FETs and thermionic valves may be used.

● Special circuits are required to match the output stages into the aerial so that maximum power is radiated.

● High-level modulation has the advantage of simplicity but requires a high-power modulation circuit and careful design of the components used, because of the high powers involved.

● Low-level modulation has the advantage of being able to produce different modulation types, such as single sideband (SSB) and independent sideband (ISB). However, the circuits following low-level modulation must work in a linear mode, with corresponding low efficiency, creating a problem with heat dissipation.

● A varactor is a device for converting voltage changes into capacitance changes. It can thus be used to vary the instantaneous value of the carrier in an FM transmitter.

● As the amplitude of a frequency-modulated wave is constant, it is possible to use maximum efficiency (class C) amplification after the modulation process has taken place in an FM transmitter.

● Pre-emphasis is used in the audio section of an FM transmitter as part of the process of increasing the signal-to-noise ratio, especially for high-frequency audio signals.

REVISION EXERCISES AND QUESTIONS

1 Calculate the frequency of oscillation for a parallel-tuned circuit with an inductance of 1 mH and a capacitance of $3\,\mu$F. How much would the frequency change if a loss of $1\,\Omega$ was introduced into the circuit?

2 Draw the block diagram of an AM transmitter, and explain the function of each block.

3 What do you understand by the term 'piezoelectric effect' as applied to a quartz crystal?

4 List the methods used to ensure that the master oscillator in a radio transmitter remains stable.

5 Explain what is meant by 'low-level modulation' as applied to an AM transmitter. List the advantages of using this type of modulation.

6 Discuss the disadvantages of high-level modulation. What are the essential features of this type of modulation in an AM transmitter?

7 Draw the block diagram of a FM transmitter. Explain the functions of the various stages.

8 What is the purpose of pre-emphasis in the AF stage of an FM transmitter?

9 How can a varactor be used to frequency modulate an oscillator?

10 Discuss possible problems that are associated with the power output stage of a transmitter.

11 What is the function of the circuit between the power output stage of a transmitter and the aerial?

12 Try to visit a broadcast radio or TV transmitter site. Can you identify the stages of the transmitter that have been mentioned in this chapter? Note the final stages of the transmitter, and the cooling methods used to prevent the valves overheating. Compare the broadcast transmitter with a small portable UHF transmitter such as that in a mobile phone. Try to identify the components that have been mentioned in this chapter.

13 Obtain a circuit diagram of a suitable transmitter. Identify the different stages, and draw a block diagram to represent the circuit diagram.

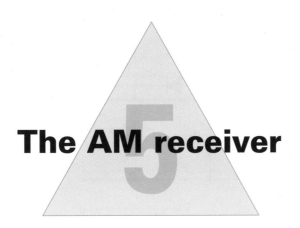

The AM receiver

Introduction

The AM receiver is an essential part of the communications network. It is probably the device that most people are interested in, as it is what they see in their homes. This chapter will deal with the receiver in its basic block diagram form. The next chapter will deal with actual circuits found in modern receiving communications and broadcast receivers.

Throughout this century, the development of the receiver has been governed by knowledge of receiving techniques and by the development of active amplification devices. Early receivers – called crystal sets – had no active device associated with them. They were then refined using the amplification that was made possible by the use of active devices such as thermionic valves. The development of the thermionic valve also led to the use of heterodyning – the mixing of two frequencies to give a difference frequency so that the frequency present at the aerial could be converted to an AF signal suitable for aural reception.

Further refinements and developments led to the receiver circuit that is normally used today. This is the superheterodyne receiver (superhet), in which hetero-

dyning is used to produce a frequency change, and the main amplification takes place at an intermediate frequency. Receiver design may now have come its full circle thanks to new technology: considerable interest is being shown in using integrated circuitry to develop the original tuned radio frequency receiving technique.

The tuned radio frequency (TRF) receiver

Figure 5.1 shows the basic requirements for a receiver. The aerial collects RF energy in the form of minute current variations across the RF spectrum. A device is then required to translate these current variations into voltage variations at one frequency or band of frequencies, which represents the band of frequencies containing the required information. If possible, this device should supply sufficient voltage to operate the next block, which extracts the information from the carrier frequency. The final block transforms the information – which could be an audio signal – into sound waves.

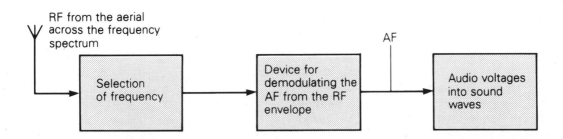

Figure 5.1 The basic requirements for a receiver.

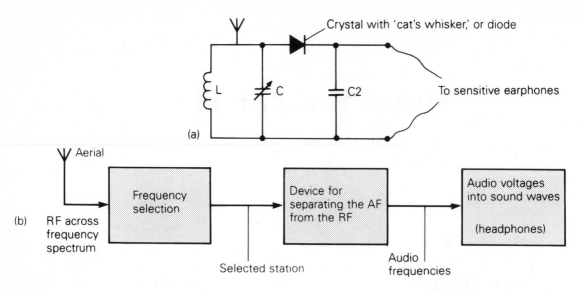

Figure 5.2 An early receiver: (a) a crystal set; (b) block diagram of crystal set.

The first receivers: crystal sets

Figure 5.2 shows an early receiver, called a **crystal set** because of the device used to extract the information from the carrier. This was originally a slice (piece) of crystal, with a spring of wire called a 'cat's whisker' making contact with its face. Connected in this way, the slice of crystal behaves as a diode, which allows current to flow through a circuit in one direction only. This is a necessary requirement for extracting information from the carrier wave.

In the crystal set shown, small currents are induced in the aerial, and flow to earth via the parallel-tuned circuit as shown. At a frequency chosen by the components in the tuned circuit (inductance L and variable capacitance C), and given by

$$f_{\text{chosen}} = \frac{1}{2\pi\sqrt{LC}}$$

a maximum voltage will appear across the face of either of the two components in the parallel-tuned circuit, giving a voltage amplification at the chosen frequency. In this way the circuit 'selects' the required frequency. The crystal and cat's whisker acting as a diode, together with the capacitor C2 and the action of the sensitive earphone, extract the audio signal from the signal received by the aerial and parallel-tuned circuit. By changing the value of the capacitance C, the receiver can be tuned to another frequency. To operate the crystal set, one had be able to select with the cat's whisker the ideal place on the crystal for the diode action to work correctly.

If we relate the crystal set to the original block diagram,

we see that the selection of frequency, and the translation of the small energy in the aerial to a voltage, is accomplished by the parallel-tuned circuit. The crystal or diode, together with C2 and the sensitive earphones, extract the audio signal: this is a demodulation or detection process.

The earphones change the variation of audio voltage into sound waves. Devices that change one form of energy into another are called **transducers**. Thus the earphone is an example of a transducer.

The developed TRF receiver

The crystal set was reasonably effective in its day, and very economical to run. However, it relied on the aerial's receiving sufficient energy to be translated into a voltage capable of operating the diode or detection circuit. The development of active components, which provided amplification, meant that weak signals at the aerial could now be selected and transferred to the detector circuit. The result was the TRF receiver as we now know it. Figure 5.3 shows the block diagram of a typical TRF receiver.

RF amplifier

This amplifies the required frequency or band of frequencies, and rejects all others. To do this it needs to be selective, and its ability to do this is referred to as **selectivity** (Figure 5.4). RF amplifiers use parallel-tuned

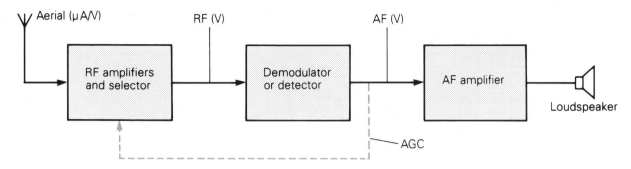

Figure 5.3 Block diagram of a typical TRF receiver.

circuits as their loads, and the amplification of the stage largely depends on the selectivity of the tuned parallel circuit. Maximum gain is governed by the parallel resonance of the tuned circuit used as the load. (There is a description of an RF amplifier circuit in the next chapter.)

The RF amplifier has to provide sufficient signal (perhaps in the order of volts) for detection (demodulation) to take place. As the input from the aerial is likely to be in the region of microvolts (one millionth of a volt, μV), the gain of the RF amplifier must be in the order of 1 000 000. This generally means that more than one RF amplifier is required, and in fact there are usually three or more separate stages.

As the receiver is designed to operate on the weakest signal, and the RF amplifier will be designed to work in class A to avoid distortion, some form of gain control will have to be incorporated to prevent overloading of these stages, and the resulting distortion, when a strong signal is received. Too much gain in this stage may also overload the following circuits (the detector and AF stages).

To overcome these problems an **automatic gain control** (AGC) may be provided. This feeds back a portion of the d.c. level available at the detection stage to control the gain of the RF amplifiers. The larger the

output from the detector, the more voltage is fed back to control the gain, thus reducing the gain of the RF stages. This is referred to as a **negative feedback loop**.

As a number of RF amplifiers are required, each one must have a load that is tuned to the same required frequency. This is normally accomplished using a system referred to as **ganging**, with all the capacitors of each load circuit being controlled by a single control.

Wide frequency bandpass circuits are possible by varying slightly the frequencies of the tuned circuits that each of the RF amplifiers loads into, and either undercoupling (loosely coupling) or overcoupling (tightly coupling) transformer circuits between the amplifiers.

Detector or demodulating stage

This stage extracts the intelligence that is being carried by the AM on the RF wave. It usually does this by rectifying the output from the RF amplifiers, and then filtering the unwanted RF frequencies and the d.c. component.

Figure 5.5 shows a typical detector circuit. It uses a semiconductor diode for rectification, working into a load R, which is usually a volume control. Capacitor Cl is the reservoir capacitor, filtering out the RF component, with C2 and R1 filtering out any remaining ripple. Note that the time constant for C2R1 should be long compared with the period of the carrier wave, but short compared with the highest audio signal period. C3 blocks the d.c. component after the volume control. Note that as the d.c. component is directly proportional to the strength of the incoming signal, it can be used as an AGC voltage. Figure 5.6 shows the detector waveforms in more detail.

The detector requires signals of the order of volts to operate if silicon diodes are used. They can be lower than this if germanium diodes are used.

Detectors can be designed that provide amplification, in which case the input to the detector stage can be made smaller.

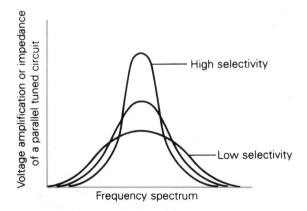

Figure 5.4 Selectivity.

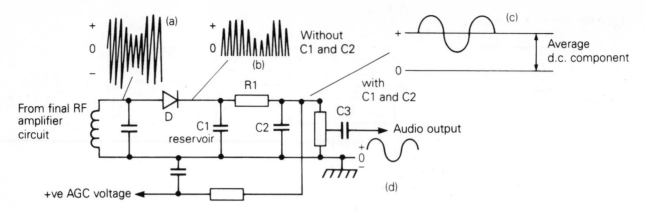

Figure 5.5 A typical diode detection circuit, showing the waveforms that are present.

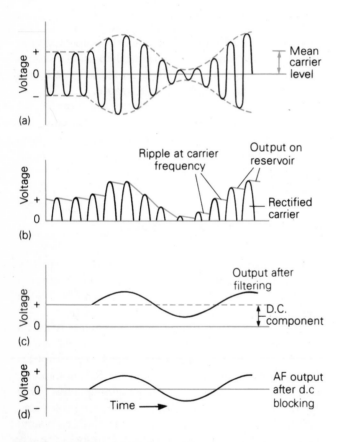

Figure 5.6 The detector waveforms in more detail.

AF amplifier

These stages amplify the output from the detector sufficiently to operate a transducer: that is, a loudspeaker. The usual combination is a class A amplifier followed by a power stage, which is normally a class B push–pull stage. Some form of volume control is normally incorporated if it is not part of the detector circuit.

Advantages of the TRF receiver

1. It is reasonably simple, and therefore conventional circuitry can be used.
2. The noise introduced by the TRF circuitry is considerably smaller than that introduced in the superhet receiver (see below). This gives a better signal-to-noise ratio and therefore a better noise factor.

Note that a fixed-tuning TRF receiver is incorporated in the more complicated superhet and called an **IF amplifier**, as will be explained later.

Disadvantages of the TRF receiver

Note that these usually refer to the tunable TRF receiver.

1. Total amplification of the RF stages (and thus the selectivity) is limited, because of instability problems. All the RF circuits are tuned to the same frequency, and so if the RF stages have a gain of 10^6, a feedback of 10^{-6} in the correct phase (**positive feedback**) would cause oscillations to build up. This is especially true of tuneable RF stages, because it becomes increasingly difficult to ensure adequate screening (to prevent positive feedback) at various frequencies. The usual practice is to screen the associated tuned circuit components with earthed metal screen cans, thus preventing any inductive interference between the various circuits.
2. The ganging of the capacitors used to tune the RF circuits is limited. This makes it difficult to control all the tuned circuits simultaneously.
3. The gain of the RF stages will vary with frequency, because the dynamic impedance offered by the tuned circuits will change as the capacitance of the circuit is altered to change the frequency. This means that the selectivity of the receiver will also change. The gain of

the stage is proportional to the dynamic impedance of the tuned circuit Z_d, which is given by

$$Z_d = \frac{L}{CR}$$

where L is the inductance of the circuit, C is the capacitance, and R is the resistance associated with the circuit. To change frequency it is usual to alter the value of C. That is:

$$f = \frac{1}{2\pi\sqrt{LC}}$$

as this is the component that gives the greatest variation that can be achieved quickly. This means that for a rise in frequency the capacitance must fall. Therefore the value of Z_d will rise, causing a corresponding increase in the gain of the stage. Thus the gain and selectivity will increase with an increase in the frequency being received.

The superheterodyne receiver (superhet)

This type of receiver attempts to overcome the disadvantages of the TRF receiver by changing the frequency of the incoming required signal to one that is fixed at a suitable frequency for amplification, and then feeding this signal into what could be termed a fixed tuned TRF receiver. This technique became possible with the discovery of a circuit that, given two input frequencies, produces outputs that include the sum and

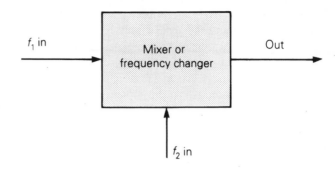

Figure 5.7 Principle of the mixer or frequency changer. The signal out consists of $f_1 + f_2$ and $f_1 - f_2$. In some mixers other frequencies are present at the output.

difference of those frequencies. This circuit is called a **mixer** or **frequency changer** (Figure 5.7).

As can be seen from Figure 5.7, the useful output from the mixer or frequency changer consists of the sum and difference frequencies that are present in the output. For the superhet principle it is normally the difference frequency that is used. The sum is used in other contexts, which are mentioned in Chapter 8. Mixers may give other outputs related to their inputs. Again, these can be put to other uses, the required output being filtered from the unwanted ones.

Figure 5.8 shows the basic block diagram of a superhet receiver.

RF amplifier

This stage is sometimes referred to as the **signal frequency amplifier**. It selects the required signal frequency and gives a degree of amplification to pass on

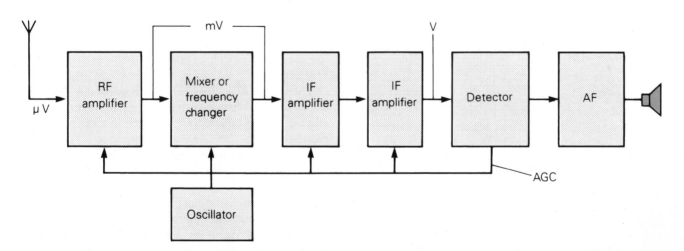

Figure 5.8 Basic block diagram of a superhet receiver.

to the mixer stage. In some commercial broadcast sets this stage is omitted, and the aerial circuit feeds straight into the mixer. The tuning of this stage is usually ganged together with the tuning of the local oscillator.

The inclusion of the RF or signal frequency stage gives the following advantages to the superhet receiver. These will be explained in more detail later in the chapter.

1. It improves selectivity and sensitivity, thus improving **image channel rejection** and **IF breakthrough rejection**.
2. It improves the noise factor (signal-to-noise ratio) of the receiver.
3. It improves the isolation between the aerial circuit and the local oscillator (LO). This avoids the phenomenon of a strong signal's altering the LO frequency slightly (**LO pulling**). It also prevents the LO from re-radiating via internal coupling to the receiver aerial.

Mixer or frequency changer

This stage mixes an input from the local oscillator with an input from the RF amplifier to give a new frequency, the **intermediate frequency** (IF). The output from the mixer can contain $f_s + f_{lo}$, $f_{lo} - f_s$, or $f_s - f_{lo}$, plus possibly other frequencies related to f_s and f_{lo} depending on the type of mixer used (f_s = signal frequency; f_{lo} = local oscillator frequency). The output tuned circuit will decide which of these frequencies is chosen for the IF.

The mixer represents the stage that introduces the most noise into the receiver. It is therefore the main factor that determines the noise factor of the receiver.

There are two types of mixing that are commonly used in superhet receivers: **additive mixing**, normally associated with bipolar transistors, and **multiplicative mixing**, normally associated with field effect transistors (FETs). These will be discussed in the next chapter.

Local oscillator (LO)

This stage provides the oscillating frequency to mix with the required signal frequency. Variation of the LO frequency provides a means by which the frequency of reception can be changed to the IF. For example:

$$f_s - f_{lo} = \text{IF}$$

assuming the output from the mixer is tuned to $f_s - f_{lo}$.

The IF is fixed, so that the output from the mixer is fixed. If the LO is tuned the only variable will be the signal frequency. Thus

$$f_s = \text{IF} + f_{lo}$$

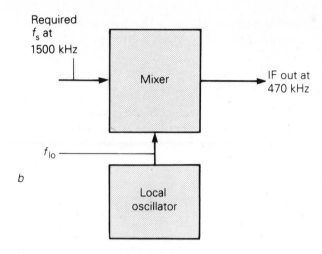

Figure 5.9 An example of the mixing process.

If the output from the mixer is $f_{lo} - f_s$ then

$$f_s = f_{lo} - \text{IF}$$

EXAMPLE

A receiver with an IF of 470 kHz is required to be tuned to a station on 1500 kHz. What frequency should the local oscillator be tuned to, assuming that it is tuned above the signal frequency?

Figure 5.9 shows the block diagram for this example.

$$f_s = f_{lo} - \text{IF}$$

Therefore

$$f_{lo} = f_s + \text{IF}$$

and

$$f_{lo} = 1500 + 470 \text{ kHz} = 1970 \text{ kHz}$$

The local oscillator in the example will have to be tuned to 1970 kHz.

The tuning capacitor of the LO is ganged with that of the RF stage to ensure correct alignment: that is, the RF and LO tuning capacitors are on the same shaft. This can create problems, however, as the two circuits are tuned over different frequency ranges, making the requirement that $f_s - f_{lo} = \text{IF}$ difficult to achieve over the frequency range.

The LO should be readily tunable but, once tuned, should be stable in operation, to provide an output that does not vary from the selected frequency. In more sophisticated receivers, a buffer amplifier is placed between the LO and the mixer.

Intermediate amplifier stages (IF)

This part of the receiver provides most of the gain. As its tuning is fixed, precautions can be taken against instability, while also giving opportunities for increasing the receiver's functions to include such things as varying bandwidth control, single-sideband operation, and auxiliary AGC.

Most receivers have at least two stages of IF amplification to achieve sufficient gain. These stages are basically fixed tuned radio frequency amplifiers.

The choice of preferred frequency for the IF is something of a compromise, and this will be discussed later in the chapter.

Detector or demodulator stage

This is normally a diode semiconductor device, possibly using germanium. This stage demodulates the carrier frequency. Remember that this is now the IF. The signal frequency has been converted to the IF in the mixer. The circuit would be the same as that described for the TRF receiver.

Automatic gain control (AGC)

This can be taken as mentioned earlier in the chapter: that is, from the d.c. level that is present at the detector stage. However, it is quite common for the AGC to have its own detector circuit, which takes a sample of the IF and feeds back d.c. to the previous IF and RF stages to control the gain.

AF amplifier

This amplifies the output from the detector to provide sufficient power to drive the transducer (loudspeaker). The same considerations apply here as to the TRF receiver discussed previously.

Note that the stages following the mixer stage are the same as for the TRF receiver, except that the RF stage is fixed tuned and is called the **intermediate amplifier**.

Advantages of the superhet over a variable TRF receiver

1. Gain and selectivity are controlled largely by a fixed-frequency IF amplifier, and hence are constant.

2. Adjustment and alignment of a large number of variable tuned circuits are avoided.
3. Stability is good, because the concentration of gain is at a fixed frequency, fairly low-frequency IF amplifier, where problems of induction and electric fields can be minimised.
4. Features such as automatic frequency control, variable bandpass circuits, crystal filters (special restricted bandpass circuits using the properties of a high-Q crystal circuit), single and independent sideband operation are possible.

Disadvantages of the superhet over a variable TRF receiver

This is a brief overview. Details of the disadvantages, and the methods used to overcome them, are described below.

1. The superhet is more noisy that the equivalent TRF receiver. This extra noise is almost all due to the presence of the mixer or frequency changer circuit, which introduces noise in the mixing process.
2. Unwanted interference may be introduced from image channel interference (sometimes referred to as second channel interference) and IF breakthrough. Note that these interfering stations will interfere only if they are being transmitted on the appropriate interfering frequency.
3. Design difficulties may be increased because of the need to keep two different frequency circuits aligned: the signal frequency circuit and the local oscillator circuit. This alignment is referred to as **tracking**.

Noise in the superhet receiver

This is a serious problem. As mentioned above, it is caused mainly by the mixing process in the mixer or frequency changer. As noise is introduced at this early stage, the signal-to-noise ratio cannot be improved after this stage. The receiver will therefore have difficulty in receiving weak signals, as they will be lost in the noise created by the mixing process.

This gives one good reason for the use of the signal frequency amplifier, which should be of a low noise type (see next chapter). In this way the signal from the aerial is amplified before entering the noisy stage, providing a better signal-to-noise ratio as well as helping in other ways (see below).

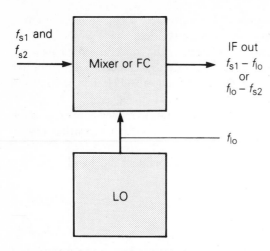

Figure 5.10 The possibility of image or second channel interference.

Interference

Image channel or second channel interference

When two frequencies are combined in the mixer or frequency changer, frequency components are produced that can be represented by $f_1 - f_2$ and $f_2 - f_1$. Both these frequency combinations can lead to the same output. This can present a problem in a superhet, where two signal frequencies can achieve the IF.

In Figure 5.10 this is represented by $f_{s1} - f_{lo}$ and $f_{lo} - f_{s2}$. Both signal frequencies, if present, would give the required IF. The IF is fixed in a superhet; assuming that the LO is tuned above the required signal frequency (as is normal in modern superhet design), there will now be two frequencies that can achieve the IF:

1. the required signal frequency of $f_{lo} - f_s = $ IF; and
2. an unwanted signal frequency of $f_i - f_{lo} = $ IF.

f_s represents the wanted signal. f_i represents the **image channel** or **second channel**, which is spaced by twice the IF away from the wanted or signal frequency.

Remember that the image channel will be present as an interfering signal only if there is a station on that image frequency (that is, twice the IF away from the wanted signal).

Here are some examples of signal frequency tuning and image channel interference frequencies:

1. Signal frequency (or frequency that the receiver is tuned to) = 4500 kHz
IF = 450 kHz
LO tuned above the signal frequency by +IF = 4950 kHz
Image channel on $(2 \times$ IF$) + f_s = (2 \times 450) + 4500 = $ 5400 kHz
2. Signal frequency (or frequency that the receiver is tuned to) = 990 kHz
IF = 470 kHz
LO tuned above the signal frequency by +IF = 1460 kHz
Image channel on $(2 \times$ IF$) + f_s = (2 \times 470) + 990 = $ 1930 kHz
3. Signal frequency (or frequency that the receiver is tuned to) = 1974 kHz
IF = 1000 kHz
LO tuned above the signal frequency by +IF = 2974 kHz
Image channel on $(2 \times$ IF$) + f_s = (2 \times 1000) + 1974 = $ 3974 kHz

In each of the above examples there would have to be a signal on the image channel frequency for interference to be present at the IF and therefore in the receiver.

However, if there *is* a signal present on the image channel, the only possible way of stopping this from passing into the receiver is to make the signal frequency stage of the superhet selective enough to reject the image channel. As we can see from the above examples, this is easier if the IF is made as high as possible. This then means that the image channel is further from the signal channel, making it easier to achieve selectivity in the signal tuned circuits. Figure 5.11 illustrates image channel rejection by the signal frequency selectivity circuits.

> For image channel rejection, the IF should be made as high as possible, so that it is easier to obtain selectivity of the signal frequency stages.

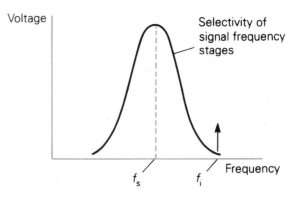

Figure 5.11 Image channel rejection. The further away f_i is from f_s, the easier it is to select f_s.

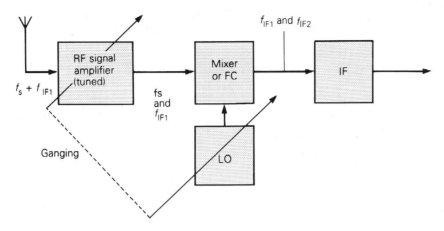

Figure 5.12 IF breakthrough.

IF breakthrough

If there is a strong station on the same frequency as the IF in the superhet receiver, there is a possibility that the station will 'break through' the signal frequency circuits and pass through the mixer. This will cause interference with the signal frequency that we are trying to received. Figure 5.12 shows the front end of a receiver with IF breakthrough. The use of the signal frequency amplifier will help to select against IF breakthrough, but in most IF designs the IF is chosen so as not to be at or near the frequencies of powerful broadcast stations.

Adjacent channel interference

This is interference that is near the required frequency of reception. It is not restricted to superhets; it is also present in TRF receivers.

This type of interference is a function of the receiver's effectiveness in selecting one frequency and rejecting another. In the superhet, as the main selectivity is achieved in the IF section, this must be able to accept the required signal and reject the unwanted. Obviously it will be easier to reject a signal that is 20 kHz away than to reject one that is only 3 kHz away.

For good adjacent channel rejection the IF should be made as low as possible. This is because the ratio of the wanted signal to the unwanted signal becomes easier to achieve as the IF is lowered. We can see this by considering the following example.

Take an IF that is 1000 kHz with an adjacent channel at 10 kHz away from the wanted signal.

Using the formula for percentage error:

$$\% \text{ error} = \frac{\left(\begin{array}{c}\text{True value} \\ \text{required}\end{array}\right) - \left(\begin{array}{c}\text{Unwanted} \\ \text{value}\end{array}\right)}{\text{True value}} \times 100\%$$

$$\% \text{ off tune} = \frac{1000 - 990 \ (\text{or } 1010)}{1000} \times 100\%$$

$$= 1\% \text{ off tune}$$

Now take an IF that is 100 kHz with the same adjacent channel 10 kHz away from the wanted signal. Again, using the formula for percentage error:

$$\% \text{ off tune} = \frac{100 - 90 \ (\text{or } 110)}{100} \times 100\%$$

$$= 10\% \text{ off tune}$$

In this case, using a lower IF has improved the IF selectivity by 9 per cent.

Thus for good adjacent channel rejection the IF should be kept as low as possible.

Tracking

This term is used to imply that the oscillator frequency follows – or 'tracks' – the signal frequency over the tuning range, so that $f_s - f_{lo}$ or $f_{lo} - f_s$ always equals the IF. That is, the oscillator frequency is always above or below the signal frequency by an amount equal to the IF, depending on which output from the mixer we choose.

It is not easy to achieve tracking. The variable capacitors used for tuning the signal frequency stage and the LO are normally ganged together on the same spindle or shaft, but the ratios of their travels and thus their capacitances will have different requirements.

Consider the receiver in Figure 5.13. It has a tuning range of 1605–525 kHz (highest to lowest frequency), which is a ratio of 3 : 1. If the IF is 470 kHz and the LO is tuned above the signal frequency, then the LO will have a tuning range of 2075 995 kHz, which is a ratio of 2 : 1. Air-spaced variable capacitors have been designed whose vanes are specially shaped to account for the differing ratios.

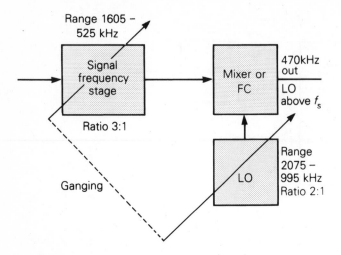

Figure 5.13 The front end of a superhet receiver, showing problems associated with tracking.

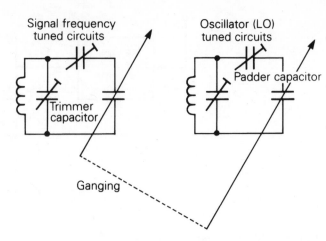

Figure 5.14 Padder and trimmer capacitances in LO and signal frequency circuits.

This can work for one range, but most receivers have a number of ranges, each with a different ratio between the highest and lowest frequencies in the signal and LO circuits. Expensive methods have been devised to solve this problem by using complicated mechanical gearing. However, the usual method of addressing the problem is to alter the value of the capacitance at the top and bottom ends of its travel by using an extra pretuned capacitor in series, called a **padder capacitor**, and one in parallel, called a **trimmer capacitor** (Figure 5.14).

By adjusting these capacitances in sequence, the error introduced by tracking can be reduced, while the alignment will be correct at three points on the band: hence the term **three-point tracking**. Domestic receivers do not normally employ a padding capacitor in the signal

circuit, but communication receivers using bandspread techniques may do to improve the sensitivity.

The effect of poor tracking is varying sensitivity across the range. In the case of badly adjusted three-point tracking, sensitivity would be maximum at the three points of perfect tracking, while between these points the sensitivity would be reduced.

Design of the superhet

As you can now see, the design of the superhet is a compromise. The inclusion of a signal frequency stage or stages increases the signal-to-noise ratio of the

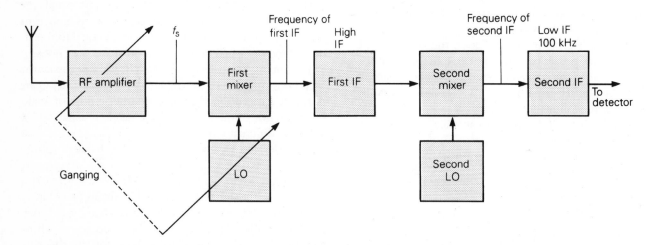

Figure 5.15 Block diagram of a double superhet receiver.

receiver, and also improves the image channel and IF breakthrough rejection because of the selective circuits used. However, these circuits have to be tuned and kept aligned with the circuits in the local oscillator, which can cause tracking difficulties.

The differing requirements of image and adjacent channel rejection together with IF breakthrough make the choice of IF difficult. Most broadcast sets having an IF between 450 and 470 kHz. Some more sophisticated superhets have two separate IFs to cater for the differing requirements of image and adjacent channel rejection. These are referred to as **double superhet receivers** (Figure 5.15).

▮ CHECK YOUR UNDERSTANDING

● TRF is the abbreviation for tuned radio frequency, usually used in TRF receivers.

● Parallel-tuned circuits can be used to give voltage amplification at the resonant frequency, given by

$$f = 1/2\pi\sqrt{LC}$$

● At this same resonant frequency the dynamic impedance

$$Z_d = L/CR$$

● The TRF receiver consists of three basic blocks: RF amplifier, detector (or demodulator), and AF amplifier.

● Selectivity is the ability of a circuit to select one or a band of frequencies and reject all others.

● The detector or demodulator stage extracts the intelligence that is carried on the RF wave.

● The superhet receiver changes the incoming carrier signal at the aerial to a suitable intermediate frequency at which amplification and adjustments are carried out.

● The device used to change the frequency to the IF is called a mixer or frequency changer.

● The signal frequency circuits of the superhet are normally ganged with those of the local oscillator.

● The main disadvantage of the superhet is the noise introduced by the mixing process.

● Interference can be introduced into the IF by the mixing process if there is a station on image or IF channels.

● For good image channel rejection the IF should be as high as possible.

● For good adjacent channel rejection the IF should be as low as possible.

● Tracking is the ability of the signal frequency tuned circuit to stay at exactly the LO frequency less the IF over the range of the band.

● Double superhet receivers are used to overcome some of the effects of image and adjacent channel interference.

1 Draw the block diagram of a tunable TRF receiver. Explain the functions of the individual stages that make up the receiver.

2 Discuss the advantages and disadvantages of a TRF receiver.

3 Draw the block diagram of the front end of a superhet receiver (that is, before the IF). Explain the function of each of the stages.

4 What are the advantages of having a signal frequency amplifier in a superhet receiver?

5 Draw a circuit diagram of a diode detector stage and explain the function of each of the components.

6 With the aid of a block diagram explain the process of receiving a broadcast using a superhet receiver. Sketch the appropriate signal after every stage, using a sine wave as the modulation on the carrier wave.

7 Explain image or second channel interference. Comment on the ways in which the effects of this interference can be minimised.

8 What is IF breakthrough in a superhet receiver? What precautions to avoid this interference should be taken in the design of the IF?

9 Adjacent channel interference occurs in TRF and superhet receivers. How can the design of the IF in a superhet help to eliminate this type of interference?

10 What do you understand by the term 'tracking' as referred to a superhet receiver? How can three-point tracking be accomplished in practice?

11 A superhet receiver is tuned to a broadcast station on 1850 kHz. If the IF is 470 kHz and the LO is tuned above the signal frequency, calculate the frequency that the LO is tuned to, and the image channel frequency if present.

12 With the aid of a manufacturer's circuit diagram of a superhet, try to identify the blocks that have been mentioned in this chapter. In the local oscillator and signal frequency stages, identify the padder and trimmer capacitor.

13 You can use the equipment shown in Figure 5.16 to observe the action of the superhet principle. Connect a signal generator to the aerial input of an AM receiver. Now connect one channel of a dual-beam oscilloscope to the local oscillator and the other to the output from the first or second IF amplifier. If a frequency counter is available, connect it to the gate output from the oscilloscope.

Tune the receiver to a suitable frequency that has no signal (say 1 MHz). Inject a signal from the signal generator at this frequency and observe the display on the oscilloscope, and the frequency recorded on the frequency counter. If the IF is 470 kHz, that

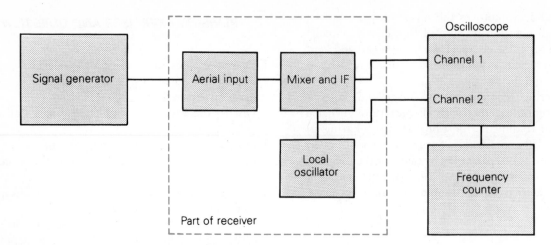

Figure 5.16 Equipment for exercise 13.

should be recorded on one trace, while the LO output should be the signal generator frequency plus the IF. What is the effect of altering the signal generator frequency slightly? What is the effect of altering the tuning of the receiver?

14 Find the AGC feedback line in the receiver and connect a high-impedance d.c. voltmeter to it. This reading is sometimes on the receiver to indicate signal strength.

Useful circuits used in AM radio reception

Introduction

In this chapter we look in more detail at the special circuits used in AM receivers mentioned in the previous chapter. We aim to show the developing nature of electronics as applied to communications.

Electronics started with the use of thermionic valves as active devices. Early receivers had designs that utilised these valves, and needed provision for power supplies to their heaters. Thermionic valves are now used only in large power-amplifying devices, such as the output from a transmitter or where a cathode ray tube is employed, as in a television. This means that the power supply circuitry can be simplified, as heater supplies for valves are not required.

Thermionic valves are normally voltage operated, but bipolar transistors are current operated. So when transistors made up the majority of active devices in the radio receiver, the circuitry had to change accordingly. However, the development of field effect transistors (which are voltage-operated devices) and integrated circuits has meant that some of the techniques used in the early valve receivers are again in use.

The development of special filters and filter circuits coupled with feedback circuits has meant that large arrays of parallel-tuned circuits in the IF with subsequent specialised tuning requirements are a thing of the past.

Signal frequency amplifiers

To achieve the benefits mentioned in Chapter 5 for the receiver, the signal amplifier must be able to transfer the small microvolt signals from the aerial into a signal at the input to the mixer in the superhet that is sufficient to achieve a reasonable signal-to-noise ratio for the receiver. As the signal-to-noise ratio becomes more of a problem at the higher frequencies, signal frequency amplifiers are more often found in VHF and UHF receivers. At the same time, signal frequency amplifiers must:

1. be sufficiently selective to reject image channel interference (twice the IF away from the wanted frequency);
2. have a rejecter circuit incorporated to reject IF breakthrough; and
3. act as a buffer stage to prevent local oscillator (LO) signals from being re-radiated from the aerial.

As the signal frequency amplifier needs to be selective (to select the required signal frequency and reject all others), we would expect to see a parallel-tuned circuit as the load of the amplifier. There is also a requirement to produce a reasonable voltage from the aerial circuit using another parallel-tuned circuit at this point. The amplifier itself will work in class A conditions so that no distortion is introduced at this stage.

Transistor signal frequency amplifier

Figure 6.1 shows the circuit diagram of a single-stage RF amplifier suitable as a signal frequency stage. Input from the aerial is selected by the parallel-tuned circuit C1L1. C1 tunes the circuit to the required frequency for reception, giving a maximum voltage output at that frequency.

The input from the aerial may also include a series-tuned circuit, tuned to the IF and connected to earth. This effectively shorts any IF at this point to earth and avoids IF breakthrough. This series-tuned circuit tuned to the IF can be inserted at different points in the signal frequency circuit as an effective **IF trap**.

L1 and L2 are mutually coupled together. This feeds the signal into the base of the transistor amplifier, working in

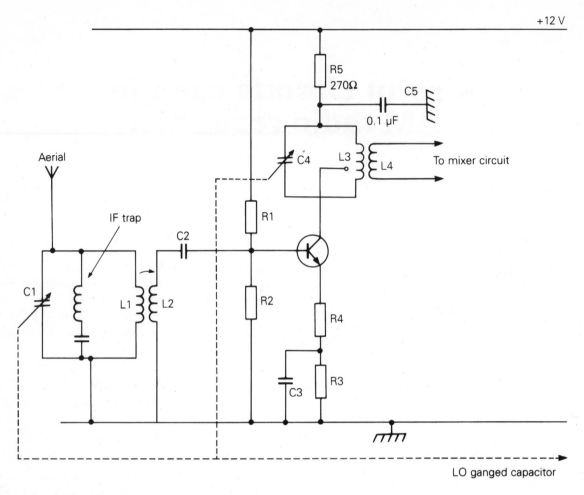

Figure 6.1 Transistorised signal frequency amplifier.

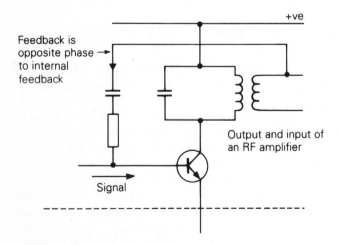

Figure 6.2 Unilaterisation circuit.

class A configuration. The load of the amplifier is again a parallel-tuned circuit, C4L3. It is tuned to the signal frequency by C4, and gives the amplifier maximum gain at that frequency. Correct loading of the transistor is achieved by tapping L3.

The output is coupled to the mixer circuit via the mutually coupled L3 and L4.

L2 feeds the base of the transistor. This is working in common emitter mode in preference to direct coupling, to avoid damping of the aerial tuned circuit, as the base input is a relatively low impedance. D.C. biasing conditions are achieved using the potential divider R1R2 together with the emitter resistors R4 and R3. R3 is decoupled by C2. R4 gives some negative feedback to the circuit, and also provides a point where AGC could be fed back. R5 and C5 provide power supply decoupling.

RF circuits using older types of transistor may be prone to degeneration of the signal, or to instability, caused by internal feedback within the transistor. This can be cured by a **unilateralisation circuit**. This consists of a CR circuit that feeds back a signal to cancel the offending feedback (Figure 6.2).

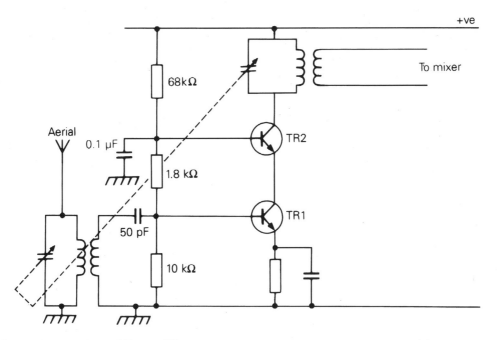

Figure 6.3 Cascode transistor RF amplifier.

Cascode transistor signal frequency amplifier

A circuit that eliminates the feedback problem is shown in Figure 6.3. This circuit is used when a wide bandwidth is required. It is used for frequencies up to 300 MHz with good isolation between the input and output circuits.

TR2 is working in the common base mode (the base is decoupled to earth), so that TR1 is working into a low impedance, giving very little gain and making the feedback possibilities negligible. However, the output impedance of TR2 is relatively high, so the output matching load can be of high impedance, giving a good overall voltage gain for the stage.

Bipolar transistors tend to be noisy devices owing to the change in carrier that is inherent in their construction and working. Field effect transistors and the integrated circuits that are now being used as active amplifying devices tend to be less noisy, as they use only one medium as a carrier.

FETs as signal frequency amplifiers

The use of FETs as RF amplifiers has another advantage. They have a high input impedance (especially if an isolated gate type is used), and so they can be connected directly into the aerial circuit without loading the aerial tuned circuit. Figure 6.4 shows single and cascode FET signal frequency amplifiers.

Because of the capacitance between gate and drain (and the Miller effect), there is a need to employ neutralisation when using a single stage. This is the equivalent of unilateralisation in the bipolar transistor. FET2 acts as a common gate amplifier, making FET1 have little gain, while obtaining voltage gain by working into a large-impedance tuned circuit (reflected via the mutual coupling).

Mixers or frequency changers

In the context of a superhet receiver, a mixer is a device used to mix two frequencies together to provide a difference frequency. When it is incorporated with an oscillator (local oscillator) it is sometimes referred to as a **frequency changer**.

The two basic methods used to produce the required mixing in broadcast receivers are as follows:

1. Two frequencies are added together, and then applied in series to a square-law device such as a diode or transistor circuit. This is called **additive mixing**.
2. Two frequencies are multiplied together in an FET or dual-gate MOSFET. This is called **multiplicative mixing**.

More sophisticated receivers sometimes use more complicated mixing circuits called **ring bridge** or **balanced modulator** circuits. These can be packaged as integrated circuits.

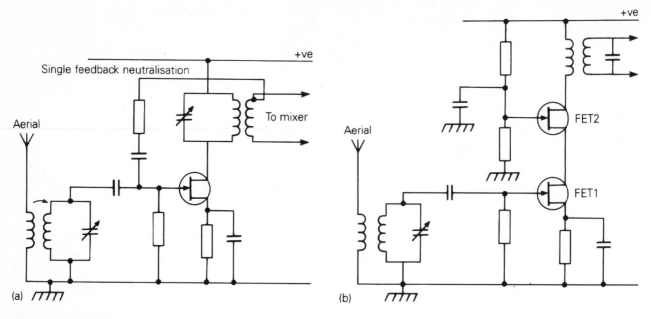

Figure 6.4 (a) Single and (b) cascode FET RF amplifiers.

Additive mixing

As mentioned above, this type of mixing uses a non-linear device. The output from a square-law device with inputs f_1 and f_2 will have the following components (Figure 6.5):

1. f_1.
2. f_2.
3. The sum and difference frequencies $f_1 + f_2$, $f_1 - f_2$, and $f_2 - f_1$.
4. Should one of f_1 or f_2 contain a range or number of frequencies – as we should expect if we represented the signal frequency coming from the aerial – these would also mix with the other frequencies to contain sum and difference frequencies that would be present at the output of the non-linear device.

5. Various harmonic frequencies of f_1, f_2, and sum and difference frequencies.

As the output to the IF from the mixer is normally the signal frequency $f_{lo} - f_s$, the output of the mixer is tuned to this frequency and to reject all other components.

As shown in Figure 6.5, the non-linear device can be a semiconductor diode working on the square-law part of its characteristics. It can also be incorporated in a transistor circuit using the base-emitter circuit, which is a *pn* or *np* junction semiconductor diode. In this way some amplification can be achieved while isolating the output circuit from the mixing process.

Figure 6.6 shows a transistor mixer with a separate local oscillator circuit. In this circuit, biasing of the transistor is such that the base-emitter operates on the non-linear part of its characteristic. The signal frequency is fed in series with the local oscillator frequency via the transformers T1 and T2. Capacitors C1 and C2 ensure that they are in series as far as a.c. is concerned. The collector circuit contains the wanted difference frequency as well as the unwanted components. This is selected by the tuned circuit C3 and L3, and transferred to the IF via transformer T3.

Frequency changer or self-oscillating mixer

Figure 6.7 shows a transistor used in a self-oscillating mixer circuit. This circuit is very popular in broadcast

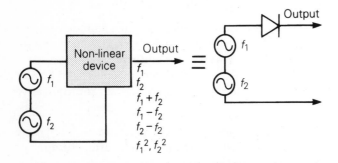

Figure 6.5 Additive mixing process.

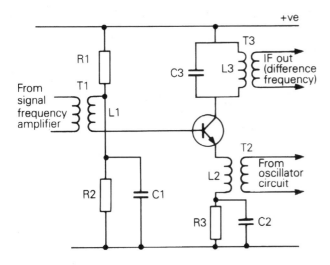

Figure 6.6 Transistor mixer circuit.

The oscillatory current circulating in this circuit is coupled to L1 and induces a voltage at this frequency into the emitter-base circuit. The transistor amplifies the oscillatory voltage and feeds it back into the tuned circuit via the mutual coupling L2L3. Mixing takes place, as in the previous circuit, when the LO and signal frequencies are in series and fed into the non-linear base emitter junction. C3 and L4 form a fixed tuned parallel circuit at the IF (difference frequency), which is fed to the IF amplifier by T3.

Multiplicative mixing

In this type of mixing, two frequencies are multiplied together inside an active device such as an FET. The output from the multiplicative mixer contains the individual input frequencies together with the sum and difference frequencies. However, it does not contain the harmonics that were present in the additive mixer.

Figure 6.8 shows the circuit of a multiplicative mixer in which the two frequencies are fed into the two separated gates, making isolation good between the two inputs and reducing the possibility of re-radiation and pulling of the LO frequency less likely. This constitutes a distinct advantage over the additive mixer. The other advantage is the larger output of difference frequency for a given input of signal frequency. This is sometimes referred to as **conversion conductance**.

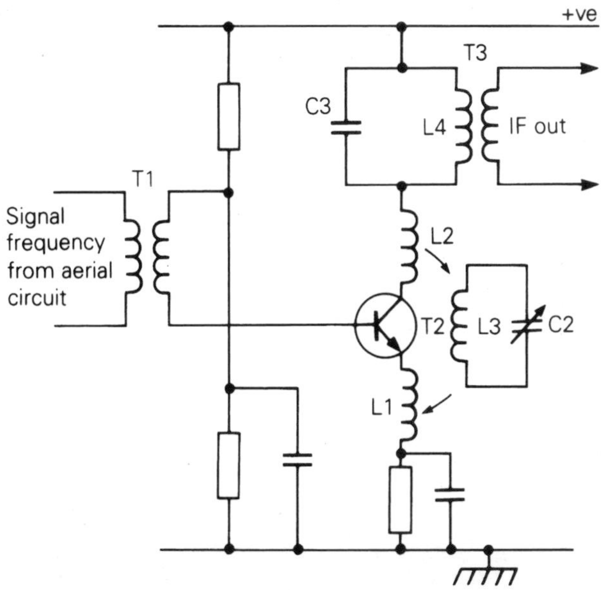

receivers because of its economy of scale and cheapness. Although it is satisfactory in performance, its faults lie in the possibility of re-radiation of the local oscillator into the aerial circuit, and the possibility of the local oscillator frequency being 'pulled' off frequency by large incoming signal frequencies.

The action of the circuit is as follows. The local oscillator section consists of the mutually coupled inductors L1, L2 and L3, the oscillatory tuned circuit L3 and variable capacitance C2, together with the transistor. L3 and C2 control the frequency of oscillation.

Figure 6.7 Frequency changer or Rheinhartz self-oscillating mixer.

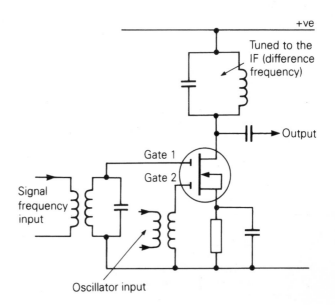

Figure 6.8 Multiplicative mixer circuit.

With the local oscillator feeding gate 1 and the signal frequency feeding gate 2, the two frequencies are effectively multiplied together. Alternatively we can consider that gate 1 alters the mutual conductance of the FET, g_m, and

$$I_d = g_m V_s$$

where V_s is gate 2 voltage. Then I_d will cause the voltage across the drain tuned circuit to vary in sympathy with the two gate voltages multiplied together. The tuned circuit in the drain is tuned to the difference frequency – the IF.

This type of mixing tends to be noisier than the additive type, because the mixing process takes place in the conduction stream inside the FET.

Local oscillator

The same stability considerations apply to the local oscillator as those discussed in Chapter 4 for master oscillators in transmitters. As there are far more receivers than transmitters, and they are far less expensive, the same strict control is not needed. Nevertheless, a local oscillator in a receiver should be designed so that stability is built in. This usually involves stabilised power supplies and perhaps a buffer load before the mixing process. There should be adequate screening between the signal frequency stages and those of the local oscillator: this is difficult, because of the shared ganged capacitance that is necessary (Figure 6.9).

In communications receivers, stability may be a prime requirement. The LO may then have crystal control for each frequency required. An oven may be used to keep the components of the LO at a constant temperature.

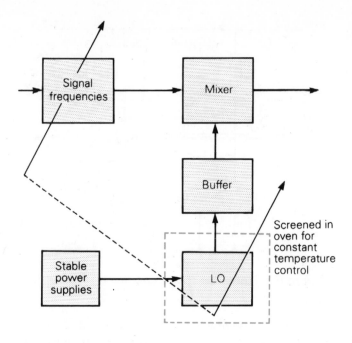

Figure 6.9 Stability requirements for local oscillator.

Other methods, such as frequency synthesis (discussed in Chapter 8), may be used.

Intermediate frequency (IF) amplifier circuits

The intermediate amplifier is a pretuned RF amplifier employed in a superhet receiver to provide most of the gain and selectivity in the receiver. As in an RF amplifier, the load of the circuit is normally that of a

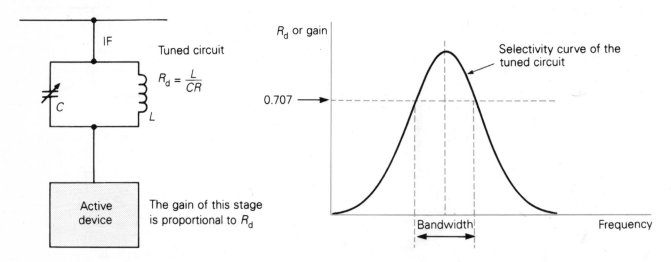

Figure 6.10 Selectivity of an IF stage.

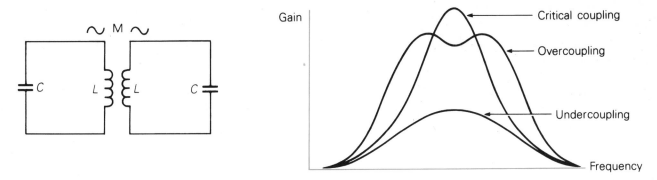

Figure 6.11 The effect of varying coupling.

parallel-tuned circuit. The selectivity of the load decides the bandwidth of the circuit, which is a function of the gain of the stage. Figure 6.10 shows how the selectivity of the parallel-tuned circuit can vary, affecting the gain of the stage and thus the bandwidth of the signal being amplified.

The bandwidth of the circuit is normally taken as the points on the curve that represent half power, or 0.707 of the maximum voltage or current gain. To increase the bandwidth of a receiver using tuned circuits, a technique called **bandpass coupling** is used. This couples a number of tuned circuits together using RF transformers. Figure 6.11 shows the effects of varying the coupling of an RF

transformer between two tuned circuits.

The required bandwidth can be achieved by varying the coupling between the tuned circuits and varying the frequencies of the tuned circuits. This is particularly important in FM and TV circuitry, where large bandwidths are required.

Figure 6.12 shows typical IF stages in an AM receiver. R1, R2, R7 and R5C1 together with R3, R4, R8 and R6C2 represent transistor biasing arrangements for TR1 and TR2 respectively. R7 and R8 give a small amount of negative feedback and, together with R5 and R6, provide a potential divider for feedback AGC to control the gain of the stages. Tuned circuits T2C3 and T3C4 provide the

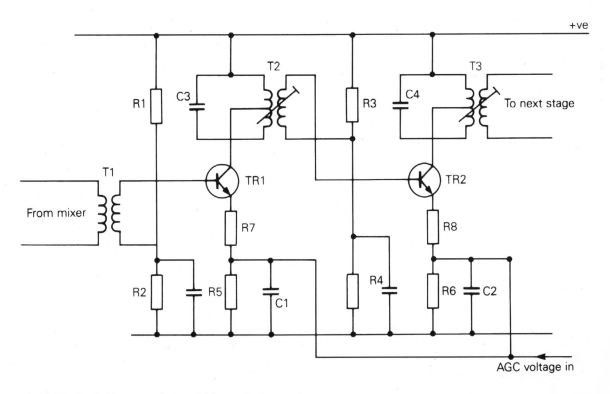

Figure 6.12 Typical IF stages in an AM receiver.

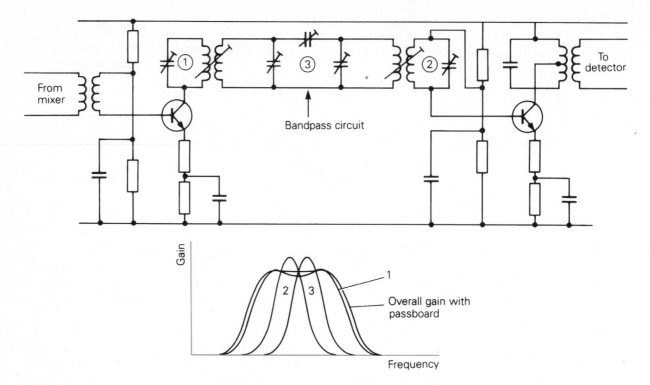

Figure 6.13 Bandpass-coupled IF circuits to give large bandpass.

load for the two transistors. Matching of the output impedance of the transistors is accomplished by tapping into the coils of the tuned circuits.

Figure 6.13 shows a typical bandpass circuit with response curves. Typically, circuits 2 and 3 may be critically coupled and tuned to suitable frequencies; circuit 1 may be overcoupled, with the tuning centred between the frequencies of 2 and 3. In this way a flat-topped wide passband can be obtained.

Some reception may require accurate small passbands to be achieved in the IF. These are normally obtained by using crystal filters, which employ the property of the quartz crystal (high Q factor) to provide an accurate narrow passband.

The use of tuned circuits in the IF section, especially when they are employed as passband circuits, requires special initial setting up. The circuits need to be tuned and accurately coupled together: hence the core adjusters indicated on the transformers, and the associated capacitors' adjustments. Special equipment is normally required, which involves a wobbulator and an oscilloscope.

> Do not adjust these circuits unless you are following the manufacturer's instructions.

To avoid the need to adjust coils and capacitances in the IF, modern circuitry may use quite complicated negative feedback circuits to replace the tuned loads mentioned above. Bandwidths are then decided by the frequency of the feedback circuits and possibly the ceramic filters that may be put into the circuit prior to the detection stage.

Figure 6.14 shows the development of an integrated circuit with active devices. The signal frequency amplifier, frequency changer and IF stages are actually on the chip with tuning circuitry connected externally. This means that the receiver can be made very small, perhaps consisting of just three integrated circuits: power supply, the integrated circuit shown above, and an AF integrated circuit.

The detector stage and automatic gain control (AGC)

The detector stage in its simplest form was discussed in the previous chapter. The purpose of the stage is to recover the information that is amplitude modulated on the carrier, which in a superhet receiver would be the IF (Figures 5.5 and 5.6). To be able to detect or demodulate weak signals it is usual to use a germanium semiconductor diode in the circuit, owing to its small forward bias characteristics (low turn-on).

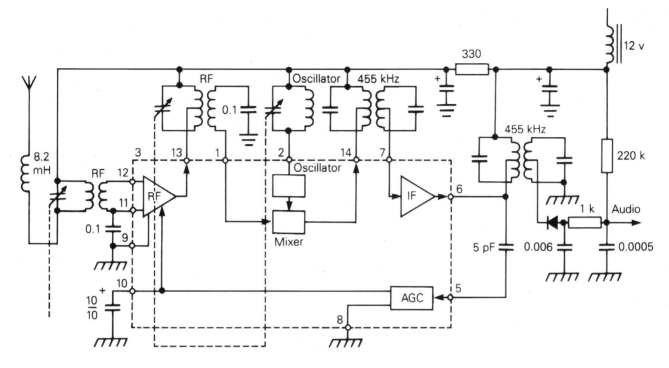

Figure 6.14 A single-chip development of the RF and IF sections of an AM receiver.

It is not possible to use this detector circuit on its own to receive continuous wave (CW) transmissions, as sometimes used in communications (Morse code). All that would be heard is a click in the output loudspeaker as the CW signal was switched on and off. It requires the inclusion of a beat frequency oscillator (BFO) in the detector circuit in series with the IF signal. The signal that it generates beats with the IF to give a difference frequency (additive mixing). The BFO should therefore be a suitable audio frequency above or below the IF. Figure 6.15 shows a detector circuit for receiving CW signals.

The reception of single-sideband signals is also a problem. It can be solved by re-inserting the carrier wave before detection takes place. In this case the BFO would be replaced by an oscillator working at exactly the IF. Figure 6.16 illustrates the modifications required for receiving single sideband signals. In more expensive communications receivers the demodulation circuit may be a balanced demodulator, possibly of the ring bridge type.

AGC

The signal strength at the aerial of the receiver may fluctuate considerably for a number of reasons to do with

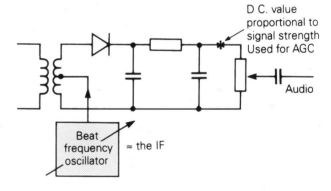

Figure 6.15 Detector circuit for receiving CW

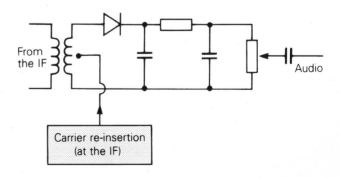

Figure 6.16 A detector circuit for single-sideband reception.

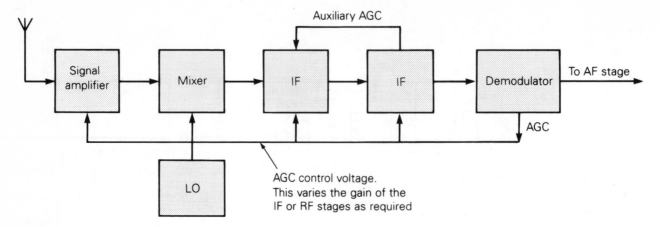

Figure 6.17 Principle of automatic gain control.

the propagation of radio waves (see Chapter 2). This would be a corresponding disadvantage in the reception of these signals, which would fluctuate wildly in sympathy with the aerial variations unless the gain of the receiver was varied to compensate.

The ideal receiver should have a more or less constant carrier level at the detector stage. To help achieve this, AGC is applied, normally to all amplifier stages prior to the detector. This ensures that the audio output from the receiver varies as a function only of the modulation that is on the carrier and not of the carrier level. AGC is also useful in ensuring that the initial stages are not overloaded with consequent distortion by strong signals, in this case by reducing the gain.

The principle of AGC is shown in Figure 6.17. The output of the detector very usefully has a d.c.

component, which is directly related to the carrier strength. Processing this d.c. to control the gains of the previous stages provides a negative feedback loop, which will ensure that the input to the detector remains fairly constant.

The action of the circuit in Figure 6.17 is as follows. When the carrier signal is strong the input to the detector will be large, making the d.c. component large. This is fed back, after filtering IF and AF components, in the correct polarity to control the bias voltages of previous stages so that their gains are reduced. This in turn will reduce the input to the detector. Conversely, if the carrier level drops then the d.c. level from the detector will drop, causing less voltage to be fed back to control the bias. The previous stage gains will increase, which will tend to keep the input to the detector constant.

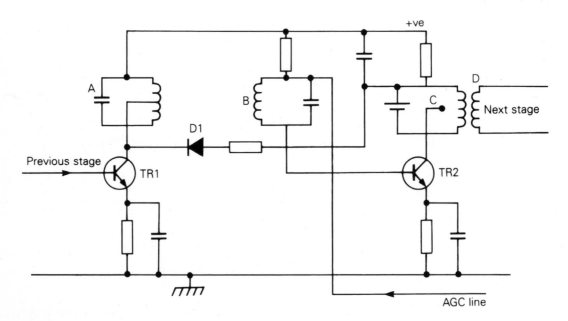

Figure 6.18 Circuit with auxiliary AGC.

Figure 6.15 shows where an AGC d.c. voltage could be taken from the detector stage.

Another form of AGC, known as **auxiliary AGC**, is shown in Figure 6.17. This is sometimes applied to give extra control and limit the amplitude of exceptionally large signals, which might otherwise overload the IF amplifier.

Figure 6.18 shows the circuit diagram of two IF stages incorporating auxiliary AGC. In this circuit, tuned circuits A and B with C and D give the required bandpass characteristics. The tappings on the coils are for matching purposes. Normal AGC is included, with voltages being fed back on this line, altering the bias applied to TR2 and therefore its gain. The auxiliary AGC effect is present using diode D1.

The auxiliary AGC action is as follows. With a strong or overload signal passing through the IF stages, the collector of TR1 will fall. This will cause D1 to conduct at a certain level, thus pegging the collector of TR1 to a limited excursion and so limiting the signal strength to the next stage.

Delayed AGC

The simple AGC described above has the disadvantage that it reduces the overall gain of the receiver when only small or weak signals are present. This is undesirable when receiving weak signals. A more satisfactory arrangement can be obtained by delaying the action of the AGC until a

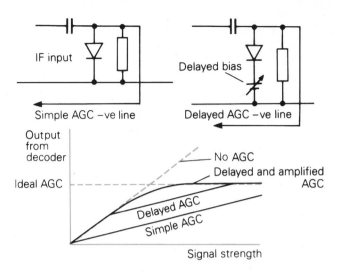

Figure 6.19 Simple and delayed AGC circuits and characteristics.

predetermined level of signal strength is reached. The AGC then becomes progressively more effective above that level (Figure 6.19).

A more effective delayed AGC can be obtained by taking the bias of the AGC diode from the first AF amplifier emitter resistor. In this way, as the signal increases so the AGC diode is allowed to conduct more, and there will be increased AGC voltage on the line. Figure 6.20 illustrates a possible circuit.

The action of the circuit is as follows. If a large signal is present AGC will operate, depending on the voltage at

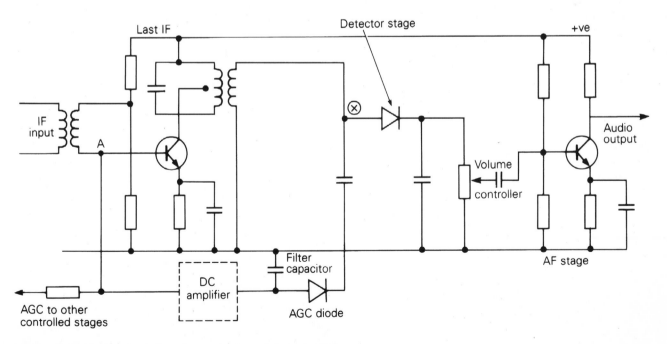

Figure 6.20 A practical circuit employing delayed AGC.

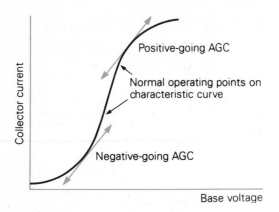

Figure 6.21 Possible operating points on the characteristic curve of an RF transistor.

point X, causing point A and other points on the AGC line to fall. The additional refinement of a d.c. amplifier can be incorporated, amplifying the AGC voltage and supplying a characteristic more like the ideal.

Note that it may not be possible to determine the direction in which the AGC voltage should go when operated; transistor amplifiers may work at the top or bottom of their characteristics. Figure 6.21 illustrates this.

Audio frequency amplifiers

A conventional audio frequency amplifier usually feeds a power stage to operate the loudspeaker. Both stages are required to work linearly: the first stage operates in class A, possibly feeding a class B push–pull stage to give the correct power to operate the loudspeaker circuit.

■ CHECK YOUR UNDERSTANDING

● Signal frequency amplifiers provide the following advantages for a superhet receiver: improved signal-to-noise ratio, image channel and IF breakthrough rejection.
● Cascode amplifiers are used to avoid the feedback problems caused by the Miller effect. They produce an effective wideband signal frequency amplifier.
● Additive mixing is the addition of two frequencies together to apply them to a non-linear device.
● Multiplicative mixing is the multiplying of two frequencies in a linear device.
● Although other frequencies are produced in the mixing process, both types of mixing produce sum and difference frequencies.
● For stable reception, it is important to have good local oscillator stability.
● IFs using bandpass coupled tuned circuits are pre-tuned, and should not be adjusted without manufacturer's literature and the correct equipment.
● To receive signals other than double sideband, special circuits are needed at the detector or demodulator.
● AGC is normally incorporated to avoid the effects of fading (reduction in signal strength) and blasting (large increase in signal strength).

REVISION EXERCISES AND QUESTIONS

1 Sketch a typical signal frequency amplifier circuit. Explain its action. What are the advantages of including this circuit in a superhet receiver?
2 Sketch the circuit of a frequency changer stage with a following IF amplifier stage, suitable for a medium-wave superheterodyne receiver. Describe the operation of these stages, and show frequencies at strategic points on the circuit if the signal frequency required is 990 kHz. Explain the reasons for your choice of local oscillator and intermediate frequencies.
3 Explain the difference between additive and multiplicative mixing. What are their relative advantages and disadvantages?
4 Give the reasons why AGC may be used in a receiver circuit. With the aid of a graph show the various types of AGC characteristics. Give a circuit of the IF and detector stages, showing how AGC is derived and applied.

The FM receiver

Introduction

This chapter looks at FM receivers in general, and examines in some detail the differences between this type of receiver and AM receivers. Commercial broadcast receivers now normally incorporate both FM and AM receivers in the same package, and this is possibly due to the similarity of most of the stages. The signal path can therefore be directed to accommodate this, being switched at points where the circuits become different.

The introduction of FM broadcasting over a number of years has meant that the advances in broadcast technology have allowed high-fidelity and possibly stereo reception to be enjoyed by a greater public. The use of FM has meant that the bandwidth of broadcast stations has increased; FM uses more bandwidth than the equivalent AM (see Chapter 3). This in turn has meant that FM stations have been moved up the frequency spectrum, and operate at VHF and UHF to accommodate the number of stations required.

The use of higher frequencies for FM has meant that the circuitry at the front end of the receiver has been modified to suit the requirements of these frequencies.

The FM receiver will generally have a signal frequency amplifier, while the front end of the receiver, comprising the signal frequency amplifier, mixer and local oscillator stages, may be put together in a unit called a **tuner**. This term is usually employed in connection with television receivers, in which the tuner is normally a sealed throwaway unit.

As the frequencies employed become higher, it may be difficult to identify conventional components easily. An inductor can be a short piece of wire, while a capacitor may be one piece of wire lying beside another, or a wire running alongside an earth point. This makes servicing of this type of circuit difficult and specialised, and also explains why tuners are usually sealed throwaway units.

Block diagram of the FM receiver

The block diagram of a typical FM receiver in Figure 7.1 shows the resemblance to the AM receiver. The main points of difference are the need for a signal frequency amplifier, the frequency of the IF, the limiter, and the

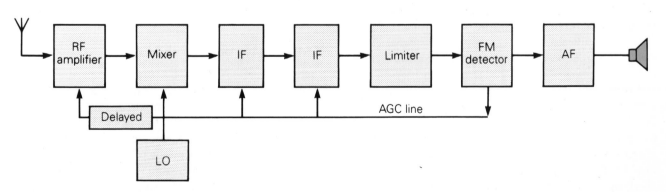

Figure 7.1 Block diagram of an FM receiver.

need for a different sort of detector or demodulator. Other differences, which do not show on a block diagram, may include:

1. specially designed AF stages to take advantage of the high-fidelity service that is available on FM;
2. special circuits to avoid the introduction of noise into the signal path (extra smoothing on the power supplies, de-emphasis circuits in the AF amplifier);
3. the possible use of muting circuits to avoid noise when tuning between stations;
4. automatic frequency control (AFC) circuits.

Signal frequency amplifier, mixer and local oscillator

These circuits serve the same purpose as those discussed in previous chapters on the AM receiver.

The problem of noise increases as the frequency of reception is increased. This, together with the fact that the IF will be increased to accommodate the extra bandwidth required for FM, will introduce more noise: so it is essential to use an SF amplifier to improve the signal-to-noise ratio.

The **signal frequency amplifier** will normally employ a wide-band cascode amplifier of the type described in Chapter 6. The use of a dual-gate MOSFET as the amplifier means that the AGC can be applied to one of the gates to control the gain of the stage without affecting the linearity of the SF stage.

Back-to-back silicon semiconducting diodes can be used to prevent overloading of the SF stage in the case of a large signal input, by connecting them in the signal input path from the aerial to earth. In this way the input to the signal frequency stage cannot rise above 0.6 V (see below).

Because of the noise problems inherent at the higher frequencies at which FM operates, care has to be taken with the **mixer** stage to prevent the introduction of unnecessary noise. Additive mixing is normally used to avoid the extra noise that might be introduced by multiplicative mixing (see Chapter 6). A strong oscillator signal (typically ten times the strongest RF signal) helps the gain of the mixer conversion and reduces the noise of the mixer noise.

Local oscillators exhibit few differences in superhet circuitry, which means that the same conditions apply as for AM receivers. In broadcast receivers it would be normal to expect a combined mixer/local oscillator circuit, as shown in Figure 6.7. However, as the IF will now be at a higher frequency, the local oscillator will also have to act at a higher frequency, which will increase the possibility of re-radiation from the aerial. In more expensive receivers the local oscillator would be separate, and possibly work into a buffer amplifier placed between it and the mixer stage.

IF stages and limiter

These now work at a much higher frequency to accommodate the higher bandwidth necessary for FM transmission. A normal FM IF would be in the region of 10.7 MHz. This tends to cause instability problems due to possible induction or radiation at these frequencies. The IF stages therefore tend to be well shielded, and screened from other parts of the circuit. Bandpass coupling in modern broadcast sets may avoid coupled circuits, and instead use pretuned ceramic or crystal boxed filters. This means that the active stage of the IF amplifier may consist of a wide-band amplifier, possibly controlled by negative feedback circuits followed by filter circuits, giving the required passband centred on 10.7 MHz.

In FM, any noise that enters the system and alters the signal will show up as changes of amplitude, such as spikes of noise. Thus to avoid noise that may be present with the signal, any changes in amplitude (amplitude modulation) need to be suppressed. This is the function of the **limiter** or limiters in an FM receiver.

Limiters are usually biased so that they operate at the top and bottom part of their output characteristics. This means that the signal will be limited by saturation and bottoming of the limiter. Figure 7.2 illustrates the principle.

The type of active device used, and the biasing applied, will determine the saturation and bottoming points that the limiter works to. In other respects the circuit will appear indistinguishable from an IF amplifier.

Back-to-back semiconductor diodes can also be effectively used as limiters (Figure 7.3). Silicon types limit the excursion of the amplitude to 0.6 V. However, with effective biasing the limiting can be preset to a known value.

FM detector or demodulator

This stage produces an alternating output voltage to supply the AF stages. The amplitude of this voltage is directly proportional to the frequency deviation of the input signal, and its frequency is equal to the rate of deviation about the mean value of the input signal.

There are a number of different circuits that can be employed for this function, using both discrete circuitry and integrated circuits. We shall look at three discrete

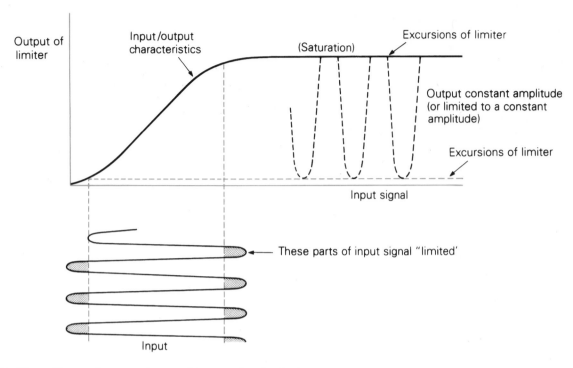

Figure 7.2 The effects of saturation and bottoming in limiters.

types in this section. They all use the principle of converting FM to AM and then rectifying the resultant waveform. We shall examine the phase-locked loop circuit later in this chapter, and look at other integrated circuit demodulators in Chapter 8.

The slope detector or demodulator

This is the easiest FM demodulator to understand, and is perhaps the basis of other types. Although it is now not used, it helps us to understand why in certain circumstances FM may be received using an AM receiver.

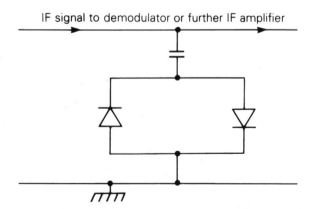

Figure 7.3 Back-to-back silicon semiconducting diodes used as limiters or SF protection against an overload signal.

The basis for this type of detector is a single tuned parallel circuit that is tuned away from the IF (carrier) by a small amount. This circuit, which is shown in Figure 7.4a, feeds a conventional AM detector. If the IF carrier is at 10.7 MHz, the tuned circuit may be tuned to approximately 10.8 MHz, placing the carrier signal halfway up the slope of the response curve of the tuned circuit (Figure 7.4b).

With the carrier frequency at rest, the gain of the tuned circuit can be regarded as average. A deviation of the carrier that is higher in frequency will cause the gain of the circuit to increase and thus the output voltage to increase. A deviation that is lower will lower the gain of the circuit, with a resultant lower output voltage (Figure 7.4c). Thus the slope of the response curve used in this way is a **frequency to amplitude converter**. The output from this circuit will still be frequency modulated, but it now has a related amplitude modulation, which follows the frequency. This can be detected using a conventional AM detector (Figure 7.5), which looks like a conventional AM-detecting circuit apart from the slight detuning of the tuned circuit L1C2.

It is now perhaps easy to see how a conventional AM receiver may be able to pick up FM signals using the technique of slightly detuning the receiver from the intermediate frequency so that the tuned circuit of the AM detector stage is working halfway up its slope.

Although the slope detector has the advantage of being

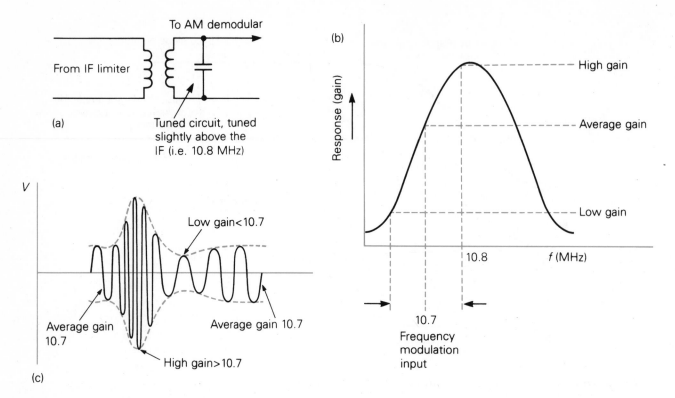

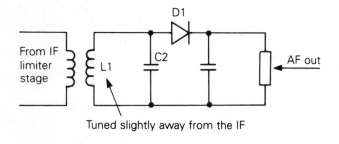

Figure 7.4 Principle of the slope detector: (a) circuit; (b) response curve; (c) resultant output, showing frequency-to-amplitude conversion.

Figure 7.5 A slope detector circuit.

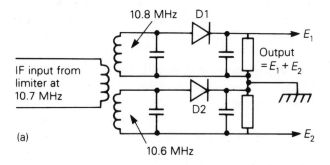

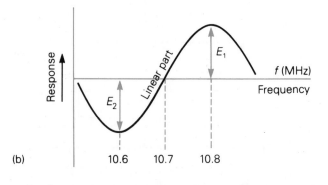

Figure 7.6 Dual-slope detector: (a) circuit; (b) response curve.

simple and cheap, its major disadvantage is that it is non-linear; only a small portion of the response curve of the tuned circuit is linear.

This disadvantage can be overcome, in theory, by using two slope detectors tuned to slightly above and below the IF carrier frequency, and taking the output across both in a circuit called a **dual-slope detector** (Figure 7.6a). In this circuit both detector outputs are added, but as the centre point of their output resistance is connected to earth, their outputs are positive and negative respectively. The outputs from each diode at 10.7 MHz are equal and therefore cancel, giving zero output. As the input signal deviation swings towards 10.8 MHz D1 conducts more than D2, giving a greater output from E_1 than E_2 and indicating

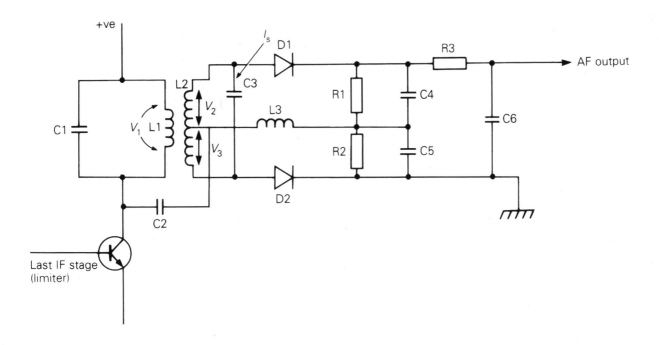

Figure 7.7 Foster–Seeley discriminator.

a positive voltage output. As the deviation swings towards 10.6 MHz then E_2 becomes greater than E_1, and the voltage output swings negative.

Figure 7.6b shows the response curve for the two circuits. It is generally referred to as the **S-curve**, with improved linearity compared with the single-slope detector.

Foster–Seeley discriminator

A further development of the slope detector resulted in the commonly used circuit shown in Figure 7.7 and called the Foster–Seeley discriminator.

The discriminator operates as follows. C1L1 act as the load of the final IF amplifier (a limiter circuit). The tuned circuit C1L1 is loosely coupled to L2C3, and both are tuned to the unmodulated carrier frequency of the IF, with sufficient bandwidth to cover the rated deviation system of the FM signal. C2 together with C4 and C5 effectively place L1 in parallel with L3. Therefore the voltage across L3 is the same as that across L1 (V_1).

At the carrier frequency (no modulation) an e.m.f. is induced in L2, and current flows in the tuned circuit L2C3. Because L2 is accurately centre tapped, equal and opposite voltages will be set up in the two halves of L2: V_2 and V_3. These voltages will be 90° shifted in phase from V_1 and of course constantly 180° from each other (as they are equal and opposite). The voltages that are

applied across the two diodes D1 (V_{D1}) and D2 (V_{D2}) are the results of phasor addition of V_2 and V_1 and V_3 and V_1 respectively. This is shown in Figure 7.8a.

Note that V_{D1} is the same magnitude as V_{D2}. Therefore the outputs across R1 and R2 will be equal and opposite in value, and will therefore cancel, giving zero output from the circuit.

When the deviation of the signal rises above the carrier frequency, the voltage across L1 (V_1) has a change of phase relationship with the induced voltage in L2, because the circuit L2C3 acts as a series circuit and becomes inductive at this higher frequency. As a result, V_2 is less than 90° while V_3 is more than 90° with respect to V_1. This causes the conditions in Fig 7.8b, when V_{D1} becomes greater than V_{D2}, causing the voltage developed across R1 to be greater than that across R2. The output voltage then becomes positive.

When the deviation of the signal falls below the carrier frequency, the circuit L2C3 become capacitive (series circuit), with the resultant phasor diagram in Figure 7.8c. This makes the output across R2 greater than that across R1, with the result that the output is now negative.

The result of the discriminator action is to have an output of changing d.c., whose amplitude is proportional to the change in frequency of the deviated carrier.

As for the dual-slope detector from which this circuit was developed, an S-curve characterises its response. The

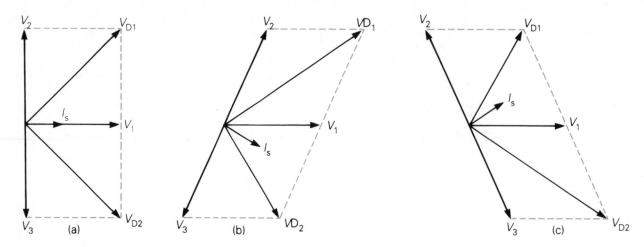

Figure 7.8 Foster–Seeley phasor diagrams: (a) no modulation (carrier frequency IF); (b) modulation above carrier frequency; (c) modulation below carrier frequency.

discriminator's operation should be kept within the 'turnover points' of this S-curve. Figure 7.9 illustrates the S-curve characteristic of the Foster–Seeley discriminator.

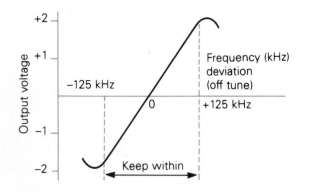

Figure 7.9 S-curve characteristic of the Foster–Seeley discriminator.

Note that the output voltage will also vary if the amplitude of the incoming IF varies. This means that two stages of limiting may be necessary. This must be regarded as one of the disadvantages of the Foster–Seeley discriminator, whose advantage is its linearity.

The ratio detector

This is possibly the most commonly used FM detector in broadcast receivers. The main reason for this is that no limiting circuit is needed, as it incorporates its own.

This circuit differs from the Foster–Seeley in the connection of the diodes (Figure 7.10). The constant input to the secondary circuit in this case is via L2, the inductive coupling to L1, and is fed to the centre tap of L3. Inductive coupling takes place between L1 and L2 but not between L2 and L3.

The outputs from D1 and D2 are therefore the phasor additions of $V_1 + V_2$ and $V_1 + V_3$. However, in this case

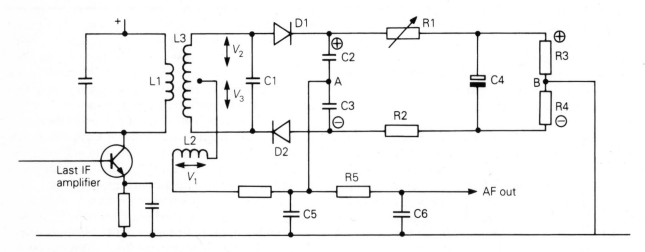

Figure 7.10 The balanced ratio detector circuit.

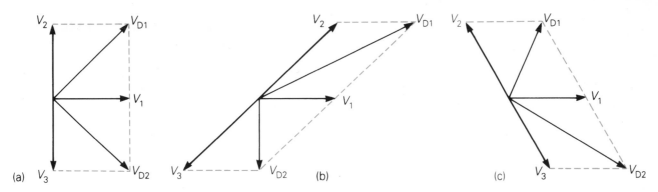

Figure 7.11 Phasor relationships for the ratio detector. (a) No output between A and B. (b) Output A goes positive with respect to B. (c) Output A goes negative with respect to B.

the diodes are connected in series addition and the polarity of the circuit is as shown. C4 is a large capacitor, which therefore has a relatively long time constant. The circuit charges up to the amplitude of the carrier wave, and thus variations in this amplitude across R3 and R4 will be removed by this circuit's acting as a type of limiter circuit.

At the carrier frequency the phasor addition is shown in Figure 7.11a. Both diodes conduct equally, which causes the voltage at points A and B to be the same (equal). As these two points represent the output from the circuit, this will cause the output to be zero.

When the FM signal deviation goes above the normal carrier frequency, the resultant phasor addition is as shown in Figure 7.11b. In this case D1 passes more current than D2, with the result that point A goes positive with respect to point B, causing the output voltage to go positive.

When the FM signal deviation goes below the normal carrier frequency, the resultant phasor addition is as shown in Figure 7.11c. In this case D2 passes more current than D1, with the result that point A goes negative with respect to point B, causing the output voltage to go negative.

The variable resistor R1 is used to balance the circuit with R2. C5, R5 and C6 form a low-pass circuit that provides de-emphasis of the AF.

The ratio detector has the advantage of combining limiting with demodulation. However, compared with the Foster–Seeley discriminator, its linearity is not so good, and the output voltage is half, owing to the directions of the diodes in the circuit.

In practice the ratio detector is used in commercial broadcast receivers, as it provides an easy answer to the problems of amplitude limiting to avoid noise; the Foster–Seeley discriminator is normally employed in more expensive receivers and in AFC control circuits.

AF and de-emphasis circuits

Figure 4.15 in Chapter 4 shows the de-emphasis characteristic that is required to take place somewhere in the audio frequency circuits of the receiver, following the pre-emphasis process that takes place in the transmission of FM.

The ratio detector shown in Figure 7.10 contains the filter circuit comprising C5, R5 and C6, whose values would be chosen to give the required 6 dB fall-off curve at low frequencies. In other circuits the precise circuits that give this required fall-off curve may be difficult to identify unless the component values are known.

Audio frequency amplifiers follow the same requirements as those for the AM receiver. They therefore require class A operation, and therefore the output stages may be class B push–pull, or class B using complementary symmetry circuitry (special AF output circuits giving a linear output). The circuitry will normally include a volume control, and possibly bass and treble controls. There is sometimes a **mute** or **squelch circuit** incorporated, which disconnects the AF amplifier when no signal is detected; this will be discussed later in the chapter.

FM compared with double-sideband AM

Advantages

1. The FM transmitter can be made more efficient than the equivalent AM transmitter. It is possible to use class C amplifiers throughout the RF part of the transmitter.

Each of the stages can work at optimum performance, as there will be no amplitude variations.

2. The FM receiver is not affected by any amplitude variations of the received signal, so external noise on the signal is reduced to a minimum, and selective fading is not present.
3. FM provides a better signal-to-noise output ratio. Noise is practically non-existent on reasonably strong signals.
4. FM exhibits capture effect: the ability of the FM receiver to receive only the stronger signals while suppressing others that may be present at the aerial on the same or near the same frequency.
5. The dynamic range of modulating signals is greater with FM, making high-fidelity and stereo reception easier to achieve.
6. There are various advantages concerning terrestrial multichannel telephony, but these are outside the scope of this book.

Disadvantages

1. The main disadvantage is the wider bandwidth required to achieve an equivalent signal-to-noise ratio; but note 3 above.
2. The coverage of FM is limited owing to propagation conditions that exist at the frequencies that it operates on – normally referred to as line-of-sight range.

Principles of automatic frequency control (AFC)

As the frequency of reception rises, the bandwidth of the signal being received becomes only a very small percentage of the carrier frequency. This means that a small percentage variation of the local oscillator will cause the selectivity of the IF circuits to reject the required signal, or will cause distortion in reception because the carrier is not exactly at the IF. This means that as the frequency used for communications goes up, so the stability of the local oscillator must be improved.

Some of this stability can be achieved by careful design of the local oscillator stage. However, this is difficult if, as is usually the case, a local oscillator needs to be tunable over a range of frequencies. The answer to this frequency stability problem can be found by the use of AFC, which is now fitted to many broadcast and most communications receivers.

Figure 7.12 shows the block diagram of a typical AFC circuit as applied to a receiver. The same circuit can equally be applied to any frequency-conscious circuit, such as one used for frequency synthesis.

In this circuit the IF is sampled through a limiter by a discriminator. The limiter in this case serves two functions: it isolates and limits any AM that may feed the discriminator. The discriminator has input circuits that are tuned to the IF of the receiver, giving a zero output from the discriminator when the IF is correct. However,

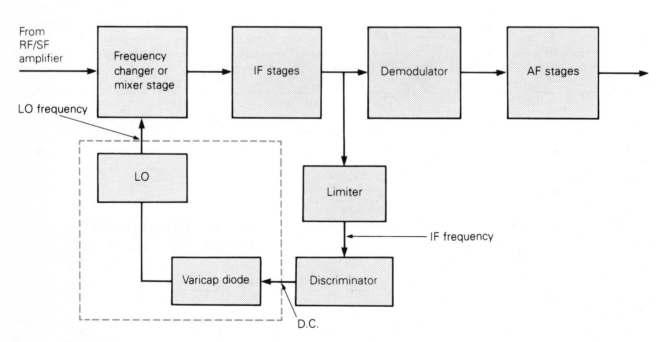

Figure 7.12 The principle of automatic frequency control (AFC) as applied to a receiver.

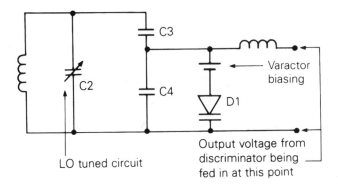

Figure 7.13 A varactor diode controlling an LO tuned circuit.

if the IF differs from the nominal value – which would happen if the receiver was mistuned or the local oscillator drifted off frequency – a direct voltage will appear at the output from the discriminator. This voltage will be proportional to the amount by which the carrier is off frequency, within certain parameters of the circuit (normally termed the **hold-in range**). This direct voltage can then be manipulated to suit the purpose of the following stage by amplification or reversing the direction of the output.

Perhaps the easiest way to control the frequency of a local oscillator circuit is to use a varactor diode. Figure 7.13 shows a possible circuit, which would be controlled by the output from the discriminator. In this circuit, L1 and the variable tuning capacitor C2 together with C3, C4 and varactor diode D1 control the frequency of the local oscillator. The varactor diode was discussed in Chapter 4, with Figure 4.16 illustrating the principle.

In the example shown in Figure 7.13, the voltage polarity for the frequency to be reduced would be derived as follows. The frequency of the tuned circuit = $1/2\pi\sqrt{LC}$. Therefore capacitance needs to increase to

lower frequency. As the diode is virtually in parallel with C2, the diode's capacitance needs to increase. This means making the depletion layer smaller, which in turn means lowering the reverse bias voltage to the varactor. So the voltage output from the discriminator should be positive-going to oppose the reverse bias applied.

The sequence of events in the AFC loop would therefore be as follows. Suppose the LO was mistuned, so that the IF was higher than the nominal IF. The carrier in the IF would be high, causing the discriminator to give a voltage that would be positive (see the description of the action of the Foster–Seeley discriminator). This happens to be the correct polarity to increase the capacitance of the varactor diode, and so increase the capacitance of the LO tuned circuit, reducing the frequency of the local oscillator and correcting the mistuning.

ACTIVITY

Work through this AFC process with the LO being tuned slightly below the frequency required to give the correct IF. Use the thought process developed above.

Muting or squelch circuits

These circuits are incorporated in receivers to eliminate the noise caused by the high gain of the receiver between the aerial and the demodulation stage when the receiver is not tuned to a station. The name 'squelch' possibly comes from the considerable objectionable noise that is present in FM systems when no signals are present.

Figure 7.14 shows possible muting or squelch circuits.

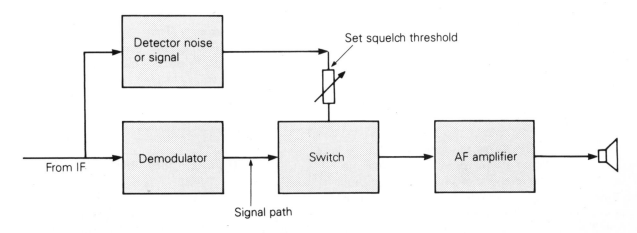

Figure 7.14 Block diagram of possible squelch circuits.

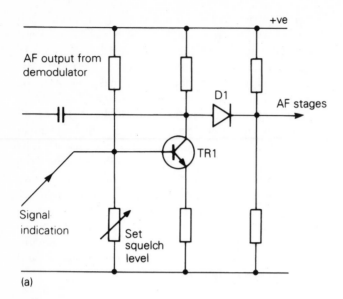

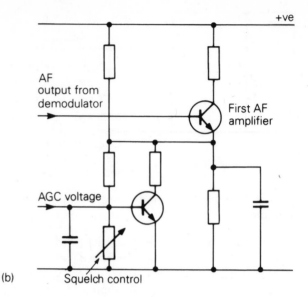

Figure 7.15 Two types of squelch or muting circuits: (a) using a diode as a switch; (b) controlling the voltage on the emitter of the first AF amplifier.

An active on/off switch is placed in the signal path of the receiver after the demodulator. It is operated by the detection of either a signal or noise, depending on the circuit used. For noise the switch (possibly a semiconductor diode) would be switched off, and for signal the switch would be switched on.

Figure 7.15 shows two types of squelch circuit. In Figure 7.15a, diode D1 is used as a switch. Under no-signal conditions, TR1 is held on by the biasing conditions set up by the squelch control; D1 is reverse biased and no signals can pass. When the carrier is present (signal), a negative voltage from an AGC-type detector cuts TR1 off; the collector of TR1 rises and switches D1 on, allowing AF to pass to subsequent stages. Note that the same circuit could use a noise detector to feed TR1 base. In this case the squelch control could be used to switch TR1 off when noise was not present (that is, signal present), and switch it on when noise made the base of TR1 positive.

Figure 7.15b shows a similar circuit, but this time controlling the voltage on the emitter of the first AF amplifier. In no-signal conditions the squelch d.c. amplifier is off, which makes the emitter of the AF amplifier high, switches that transistor off and allows no AF signal to pass. On reception of a signal the AGC line goes high, switching on the squelch d.c. amplifier. This causes its collector to fall and therefore the emitter of the AF amplifier to fall, allowing the AF signal path to be completed.

CHECK YOUR UNDERSTANDING

● FM stations operate in the VHF, UHF and higher-frequency bands.
● To increase the signal-to-noise ratio most FM receivers have signal frequency stages.
● Additive mixing is normally used in FM receivers. FM IF stages work at higher frequencies than AM to accommodate the extra bandwidth required. A common IF is 10.7 MHz.
● FM demodulators, with the exception of the ratio detector, require limiter stages to prevent AM excursions of the signal.
● Capture effect is the ability of the FM receiver to receive only the stronger of two or more signals when they are close together at the signal frequency.
● AFC is used to help the stability of the receiver by controlling the frequency of the local oscillator circuit.
● Squelch or muting circuits are found in FM receivers to prevent the reception of noise when no signal is present.

REVISION EXERCISES AND QUESTIONS

1 Using knowledge gained from AM and FM sections of this book, draw a block diagram of a receiver suitable to receive AM and FM signals. Show

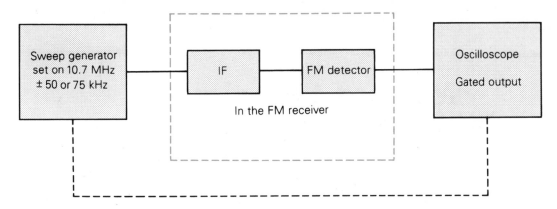

Figure 7.16 Equipment for exercise 10.

appropriate switching where necessary. Explain the main differences between the two systems.

2 Explain the process of demodulation in an FM receiver. Draw the circuit of a suitable demodulator, and explain how it works.

3 List the advantages and disadvantages of FM compared with AM.

4 Why is a signal frequency amplifier normally used in an FM receiver circuit? Why is this generally a cascode circuit?

5 Explain why additive mixing is preferred in an FM receiver.

6 It is sometimes possible to receive FM signals with an AM receiver by detuning the signal. Explain why.

7 With the aid of a block schematic diagram describe the application of automatic frequency control to an FM radio receiver. Why is AFC more necessary at VHF than at lower frequencies?

8 What does a mute or squelch circuit do? Draw a possible circuit and explain its action.

9 Obtain a suitable circuit diagram of an FM receiver. Identify the stages of the receiver, and draw a block diagram to represent them.

10 This activity observes the S curve associated with FM detectors. See Figure 7.16. Connect a sweep generator at the IF (usually 10.7 MHz), and arrange it so that it sweeps through ±50 or 75 kHz, injecting the signal into one of the IF stages. Connect an oscilloscope to the output from the detector and arrange for the timebase to trigger the output of the sweep generator. The oscilloscope trace should show the S curve that is mentioned in the chapter.

11 Identify the de-emphasis capacitor in the AF section of the receiver. If you disconnect it, you should hear noise in the form of a background 'hiss'. Reconnect it and the noise will disappear.

Modern techniques in radio communications

Introduction

Although this is an introductory book, communications have been developing so rapidly that it would not be complete without a chapter on modern radio and TV communications equipment. The treatment will be elementary, but we hope it will be sufficient to give the student a basic grasp of the subject before going on to more advanced studies.

The rapid expansion of communications in recent years has been largely due to advances in digital circuitry and the use of integrated circuits. Different techniques have had to be developed, while communication equipment has generally become much smaller, and costs have reduced. Servicing of equipment has become part of the 'throwaway culture', with faulty circuits being either completely replaced or serviced using replacement boards.

The complex nature of the new circuitry means that service engineers now have to have a 'systems' approach to their trade.

Frequency synthesis

We have already emphasised the need for stable oscillator circuitry in transmitters and receivers. While it is possible to make one oscillator stable at one frequency, it becomes increasingly difficult when a number of different discrete frequencies are required, and very difficult indeed when a continuous range of frequencies are required, as in a local oscillator circuit for a superheterodyne receiver.

We introduced AFC in the previous chapter, and this of cause helps in the stabilisation process. However, with the advent of digital circuitry it has been possible to produce, very easily, frequency divider circuits. Integrated circuits are readily available to achieve this and other functions required in the process of frequency synthesis. This has made the process easier to achieve, and it is especially useful in transceiver circuit design, where a transmitter frequency and local oscillator frequency can be obtained from the same source.

Basic principles of frequency synthesis

This circuit derives a large number of discrete frequencies, singly or simultaneously, from an accurate high-stability crystal oscillator. All the frequencies produced will have the same accuracy as the original source.

The block diagram in Figure 8.1 shows the principle of the **direct method** of frequency synthesis. A stable oscillator works at a standard frequency, in this case 5 MHz. To ensure that the standard frequency remains stable, the crystal-controlled oscillator is placed in a temperature-controlled environment. In this example the 5 MHz stable frequency is then frequency divided in stages to give outputs at 1 MHz, 100 kHz, 10 kHz and 1 kHz. These outputs are fed to frequency multiplication circuits to give f_2, f_3, f_4 and f_5. The sums of these frequencies are mixed, usually using balanced modulator circuits (to reduce the number of output frequencies) with an appropriate filter to give a final output of $f_1 + f_2 + f_3 + f_4 + f_5$. Thus, by correct switching and multiplication, continuous frequency coverage can be given in steps of 1 kHz. This direct type of frequency synthesis suffers from a number of disadvantages:

1. The mixing process may produce a number of spurious frequencies.
2. The filter circuits are many and costly.
3. The amount of switching required is excessive.

For these reasons, direct frequency synthesis is not the preferred method in modern systems.

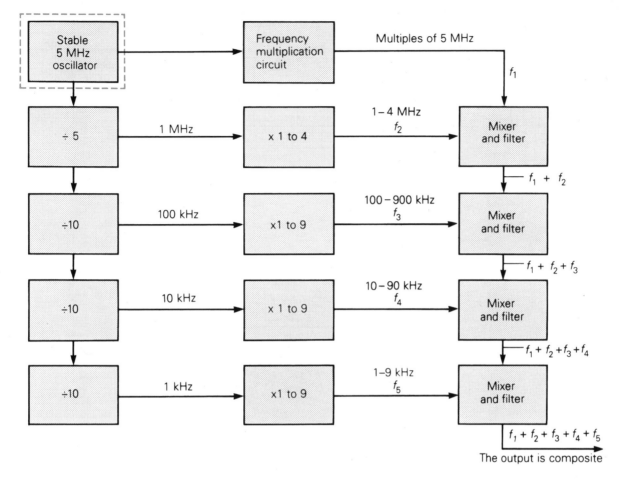

Figure 8.1 Principle of the direct method of frequency synthesis.

Indirect frequency synthesis

To overcome the disadvantages of the direct method of frequency synthesis, most modern circuits use the **indirect method**. The basic principle of this method is shown in Figure 8.2, where a standard frequency is used to control the output frequency of a voltage-controlled oscillator using a phase-locked loop. This circuit works in the same way as that described for the AFC circuit. The phase detector detects any differences in frequency

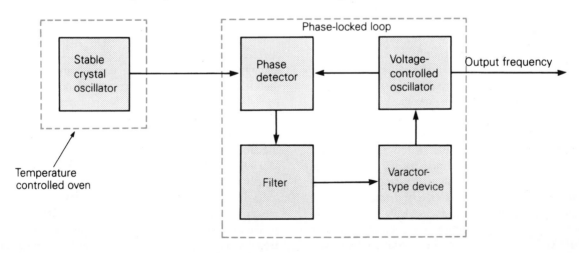

Figure 8.2 Principle of the indirect method of frequency synthesis.

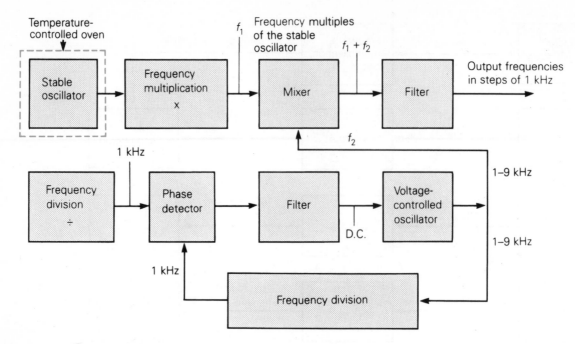

Figure 8.3 A possible frequency synthesiser using the indirect method.

between its two inputs, and produces an output voltage whose amplitude and polarity are proportional to the phase difference between the two input frequencies.

The output voltage from the phase detector is fed via a filter circuit (to remove any spurious noise) to a varactor diode, whose capacitance changes in sympathy with the output voltage from the phase detector. This controls the output frequency from the voltage-controlled oscillator, and brings the frequency back to the standard frequency derived from the stable oscillator.

In practice, many frequencies are needed from a frequency synthesiser – indeed that is the purpose of the circuit – so the above circuit is modified by the use of frequency multiplication and dividing circuits.

One possible example is shown in Figure 8.3. In this circuit both the direct and indirect methods of frequency synthesis are used, but the mixing process is kept to one stage, and so the disadvantages of the direct method are kept to a minimum. A stable temperature crystal-controlled standard oscillator works at a suitable frequency, which is multiplied to within 1 MHz of the required output frequency. This is fed to the mixer circuit (normally a balanced modulator), where it is mixed with the frequency of the voltage-controlled oscillator. The sum frequency is used and filtered. At the same time, a divider circuit produces 1 kHz for the phase detector in the phase-locked loop. One reference comes from the fixed stable oscillator and the other from the voltage-controlled oscillator, which can be set at any

frequency from 1 to 9 kHz in intervals of 1 kHz. The voltage-controlled oscillator is thus kept at the correct frequency required by the phase-locked loop.

This circuit could be used for a transceiver. In this case there would be two output frequencies, fed from two separate mixers, all controlled from the same stable oscillator.

Satellite communications

This section provides a short introduction to satellite communications, illustrating the main points. Deeper study is beyond the scope of this text.

The race to the moon that took place between the USA and Russia advanced the technology of launching satellites and the art of electronic miniaturisation in the late 1960s and 1970s to such an extent that communications satellites have now become the main carriers of long-distance communications. Satellites now represent the main source of navigation accuracy for the marine and avionics industries. They provide most international telephone traffic and almost all international and domestic long-distance television. Recent developments have meant that satellite domestic television has been made available using quite small receiver aerials. As standards are now being discussed and produced internationally to allow digital satellite broadcasting, it is confidently predicted that satellites will be broadcasting domestic digital TV in the near future; eventually all TV will go digital and be available from satellites.

The main technological development has been the ability to put large satellite payloads into the sky in precisely the required **geostationary orbit**. This is an orbit in which, to an observer on the ground, the satellite appears to remain stationary. This then allows the satellite to receive and retransmit radio transmissions to quite a large section of the earth's surface.

The increased size of satellites, together with advances in miniaturisation, has meant that the aerial size and transmitted power from the satellites have increased, making reception on the earth's surface easier. In addition, an increased ability to target 'footprints' on the earth's surface as main reception areas for the satellite's transmissions has meant that receivers do not need huge and costly aerials.

A satellite in its most basic form is really nothing more than a communications relay station in space, which has the advantage of direct line-of-sight communications and the capacity to reach most of the earth's surface, apart from the polar regions (unless the satellite is in a polar orbit).

As mentioned above, the preferred orbit for a communications satellite is one that appears stationary to someone on the earth's surface. This should be positioned over the equator, and move in a west–east direction at an altitude of 38 870 km (or 22 291 miles).

Ground linkage with the satellite is by two separate paths with a transponder (receiver and transmitter) to connect these paths in the satellite.

The **uplink** consists of the ground station, the transmission path and the satellite's receiving aerial. Because space on board the satellite is limited, its receiving aerial is made as small as possible so that its transmitting aerial can be made as large as possible. To compensate, the size and power of the ground station's transmitting aerial are increased.

The **downlink** normally refers to the satellite transmitter output power, the transmitter aerial, and the area on the ground that the transmission from the satellite will service. This is called the **footprint** of the satellite.

Most satellites are powered by solar cells, and this severely limits the amount of available power, although this has been increasing with every launch as the technology of solar energy improves. As the amount of power that can be made available for transmission is limited, the size and design of the transmitter aerial are important considerations.

The design of the aerial controls the footprint of the satellite. Figure 8.4 shows different footprints. The footprint can be made large, in which case the limited transmission power from the satellite is spread widely.

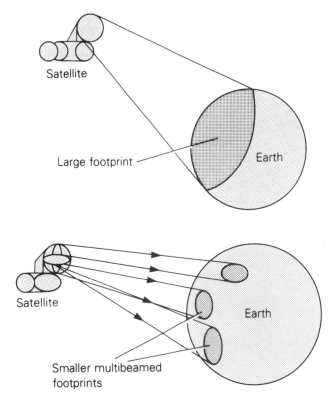

Figure 8.4 Footprints of satellite aerial beams.

Alternatively it can be channelled into a multibeam. In this case the footprint will consist of a number of small areas, each receiving a relatively stronger signal.

Coverage of the earth's surface has been achieved (apart from the polar regions) using Inmarsat satellites, and requires the use of at least three satellites. (Inmarsat is the International Maritime Satellite organisation, responsible for maritime communications, distress and safety services through communications satellites.) This system gives communications coverage for marine, aeronautical, and other vehicles over the earth's surface. It therefore uses a large footprint from each of the satellites' transmissions.

The transponder

This consists of a high-gain receiver, a converter to the down frequency, and a power amplifier to give sufficient power to the aerial. Figure 8.5 shows a typical single-channel transponder used in an Intelsat satellite. (Intelsat is the International Telecommunications Satellite organisation, responsible for the worldwide satellite communications system.)

In this particular transponder, single conversion takes

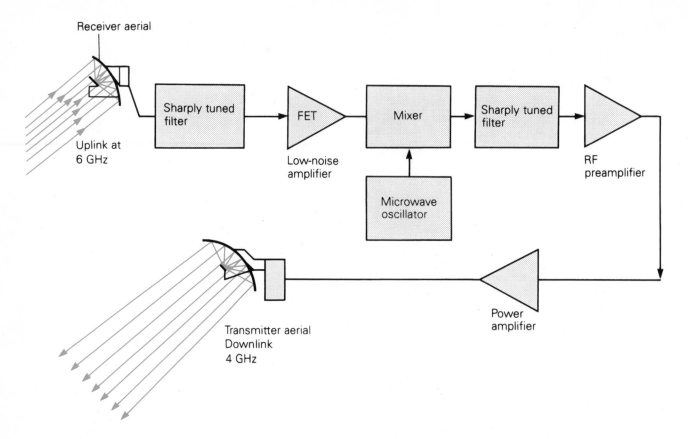

Figure 8.5 Typical block diagram of a single-channel transponder.

place from the uplink of 6 GHz to the downlink of 4 GHz; other conversions (such as 14 GHz to 11 GHz) may require a double conversion. The sharply tuned filters remove unwanted frequencies that may be present. Redundancy may be built in as back-up in case any of the active devices become faulty. This is controlled by the uplink, which can also control the gain of the transponder.

Hi-fi and stereo reception

The requirement for higher fidelity in the reception of sound was satisfied by increasing the bandwidth of the transmitted sound, and by ensuring that the receiver and associated loudspeakers were equipped and designed to receive the increased bandwidth. In the UK and Europe, high fidelity (hi-fi) was restricted to FM, because of the number of stations occupying the AM medium-frequency bands, and the improved signal-to-noise ratio of FM.

It soon became evident that the public required a more realistic form of sound, especially when listening to music. This took the form of **stereophonic broadcasting**.

Stereo (from the Greek word meaning 'solid') in this case is delivered by using two microphones, usually placed at a distance to the left and right of the required sound source. The transmitter generates and transmits both these audio signals. The receiver must be equipped to receive and separate these signals and deliver them to right-hand and left-hand loudspeakers, or a stereo headset. Figure 8.6 shows stereo broadcasting diagrammatically.

Most countries now use a system of stereo broadcasting, originally developed in the USA, called the **Zenith pilot tone system**. The main features that are desirable when broadcasting stereo are as follows:

1. The stereo signal information should be transmitted via a single RF channel without causing a significant increase in bandwidth over the corresponding mono (monaural) transmission. (The bandwidth required for mono transmissions in the UK is 240 kHz.)
2. The stereo multiplex (multichannel) system should not degrade the performance of the mono transmissions, and any degradation in the stereo system signal should be as small as possible.
3. The systems must be compatible: that is, a mono re-

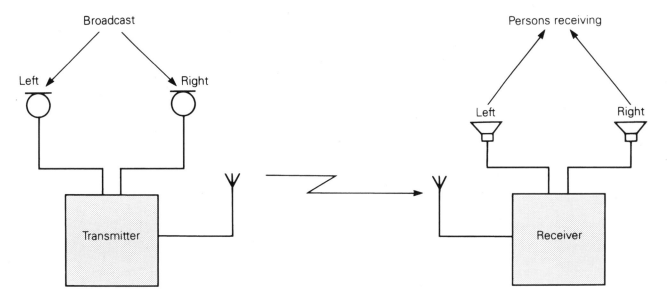

Figure 8.6 Stereo broadcasting.

ceiver should be able to receive the stereo signal as a mono signal.
4. A stereo receiver must be able to function with a non-stereo transmission: that is, produce a mono output.

Figure 8.7 shows the arrangement in the audio section of an FM stereo transmitter. Note that all the stereo circuitry is contained in the audio section. In this circuit, outputs from the left and right microphones are balanced by a resistor to earth and fed into two preamplifiers giving outputs L and R. These two outputs are mixed in two separate processes, creating outputs $(L + R)$ and $(L - R)$. $(L + R)$ is fed by a filter as a separate input into a linear adder supplying approximately 45 per cent of its output. The $(L - R)$ output channel is fed to a balanced modulator. Here it is mixed with a 38 kHz subcarrier. This gives an output that contains upper and lower sidebands only, because of the $(L - R)$ modulating waveform (Figure 8.8). (To refresh your memory, refer back to Figure 4.13 and the associated text dealing with balanced modulation.)

These VLF/LF subcarrier sidebands are fed to the linear adder, which produces approximately 45 per cent of its output from this source. The remaining 10 per cent of the output from the linear adder is supplied by an output from the 19 kHz oscillator acting as a pilot carrier for the receiver circuits. The composite output from this circuit (from the linear adder) is shown in Figure 8.8 and consists of $(L + R)$ (45 per cent), the $(L - R)$ subcarrier sidebands (45 per cent), and the pilot carrier of 19 kHz (10 per cent). These outputs are fed to the FM modulation section of the FM transmitter (see Chapter 4) prior to transmission

taking place.

As with the transmission part of the stereo broadcast, the FM receiver is conventional. All the stereo processing takes place in the audio section of the receiver. Figure 8.9 shows the arrangement to decode the stereo paths in the audio section of the receiver. In this circuit, the three separate signals that were produced in the audio section of the transmitter are separated by filter circuits. The 19 kHz pilot feeds a frequency-doubling circuit. The 38 kHz output from this circuit is used as the carrier reinsertion in the decoder. It is fed to the centre tap of the secondary winding of the transformer forming the balanced demodulating circuit with D1 and D2. It also operates an indicator light when stereo is being received.

The $(L - R)$ subcarrier sideband is fed by the primary of the transformer to the secondary, where the carrier is reinserted prior to demodulation taking place. D1 produces an output which is $+(L - R)$, while D2 produces an output which is $-(L - R)$. The $(L + R)$ circuit is fed by R_r and R_l to combine with $+(L - R)$ to give 2L and with $-(L - R)$ to give 2R respectively: that is, twice the output from each line. These outputs are fed to left and right amplifiers respectively and onwards to the left and right speakers.

To satisfy one of the original requirements, when receiving mono signals L + R must be the only signal present: the 19 kHz oscillator is switched off in the transmitter. This ensures that the balanced modulator does not function in the transmitter, giving no output of the $(L - R)$ signal or the 19 kHz pilot signal. In the receiver, $(L + R)$ will be fed to both left and right amplifiers and thus to the left and right loudspeakers.

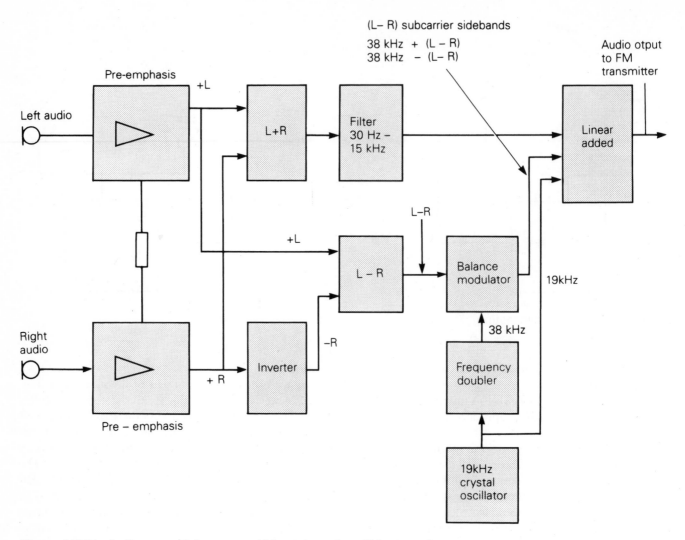

Figure 8.7 Block diagram of the stereo AM section of an FM transmitter.

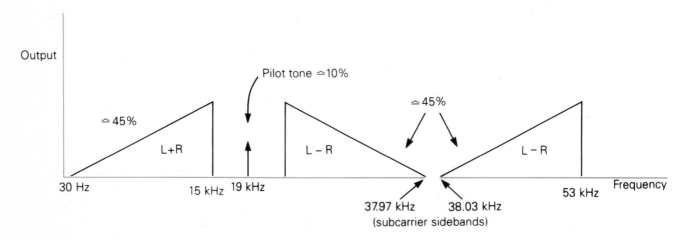

Figure 8.8 The frequency spectrum at the output from the stereo processing.

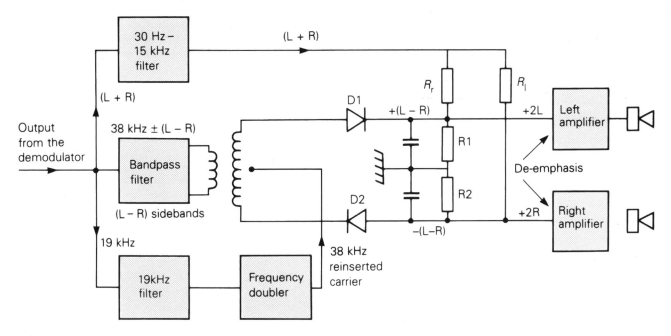

Figure 8.9 Decoding of the stereo signal in the audio section of the receiver.

The phase-locked loop detector (PLL)

This circuit provides another method of demodulating FM signals. It has distinct advantages over other methods, but it was not economically viable until it could be produced as an integrated circuit. As it becomes cheaper to produce, it will be more widely used, both in communications and in broadcast FM receivers.

The basic block circuit is shown in Figure 8.10. It consists of a phase detector, a low-pass filter, an amplifier and (to make up the loop) a voltage-controlled oscillator.

All circuitry, apart from the low-pass filter and possibly the varactor diode associated with the voltage-controlled oscillator, can be incorporated in an integrated circuit.

When the IF is constant (no FM present), the phase detector produces an output that is proportional to the difference in phase between the IF and the voltage-controlled oscillator. The resultant error output voltage is fed through the low-pass filter to eliminate any noise and via an amplifier to the input to the VCO in such a polarity as to produce a signal that reduces the frequency difference (phase) between the IF and the VCO output. Eventually the VCO output frequency will equal the IF, and the circuit is then said to be **locked**. Note that there will always remain a slight phase difference between the

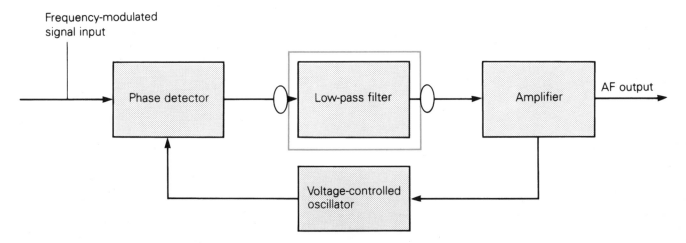

Figure 8.10 Block diagram of a phase-locked loop frequency demodulator.

IF and VCO output so that a lock can be maintained.

When the IF is frequency modulated the phase detector output voltage will vary in sympathy with the instantaneous value of the modulation and produce an output that is the required conversion of FM to AM.

The PLL offers considerable advantages as an FM demodulator over other demodulators:

1. Only an external capacitor is used to tuned the VCO to the IF; there is no tuned circuit.
2. The PLL self-adjusts within limits to match the IF.
3. It introduces very little noise or distortion.
4. It can be manufactured as a small integrated circuit with few external components.

CHECK YOUR UNDERSTANDING

- Frequency synthesis is a process that derives a number of different frequencies from one source.
- All frequencies produced by frequency synthesis have the same accuracy as the original source.
- The accuracy of the frequency synthesis process is normally derived from a stable oscillator. This is crystal controlled; its circuitry is placed in a temperature-controlled area.
- There are normally two methods of frequency synthesis: direct and indirect. The preferred method is indirect.
- Communications satellites are normally placed in a synchronous orbit around the Earth, so that to an observer on the Earth's surface the satellite appears stationary.
- A satellite in its basic form is a communications relay station.
- A satellite transmits and receives on different frequencies.
- Stereo processing takes place in audio circuits. Mono receivers must be able to receive stereo transmissions, and stereo receivers must be able to receive mono transmissions.

- Phase-locked loop detectors are becoming increasingly popular as FM demodulators, in the form of integrated circuits.
- The PLL has a number of distinct advantages over other demodulators.

REVISION EXERCISES AND QUESTIONS

1. What is meant by frequency synthesis? Why is it used in modern communications equipment?
2. Explain the difference between the direct and indirect methods of frequency synthesis. Use a block diagram to illustrate the use of the indirect method to produce a number of different frequencies for a radio transmitter or local oscillator in a receiver.
3. Explain how stereo sound is broadcast, and what provision must be made for mono broadcasts and receivers.
4. Draw a block diagram of the audio circuits required to receive stereo broadcasts. Explain the purpose behind the 19 kHz transmission.
5. What is meant by the term 'footprint' as applied to satellites?
6. What are the three main elements in the communications link using a satellite? Why is the transmission side of the satellite assigned more power than the receiving side?
7. Draw the block diagram of a typical transponder circuit in a single-channel satellite system. Explain the purpose of each block.
8. Explain how the phase-locked loop detector works in the reception of FM signals.
9. What are the main advantages of using PLL circuits in the reception of FM?
10. Obtain the circuit of a frequency synthesiser (in either a transmitter or a communications receiver). Identify the stages in the synthesiser, and draw a block diagram to illustrate them.

Principles of television

Introduction

For all kinds of broadcasting – whether for radio, television or data – a suitable carrier must be radiated from the transmitter. In any country, the carrier frequency (whether UHF or VHF) must fit within the channels chosen by that country. All channels are allocated by international agreement.

For land-based television transmissions, the video signal modulates the carrier signal using amplitude modulation. The sound signal uses frequency modulation. However, for broadcasting by satellite, frequency modulation is used.

For reception, all receivers must be capable of receiving both the above transmissions with full compatibility.

We begin in this chapter by looking at how a video signal is produced using a camera tube called a vidicon. The vidicon tube utilises photoelectric principles. We also look at the way a complete picture or scene is built up by using a system of scanning. Two types of scanning are discussed, and their relative advantages. The idea of picture synchronisation is also mentioned.

We look at overall system bandwidth with reference to the fineness or resolution of the video signal and also its reproduction rate.

We also look at different ways of modulating the video signal onto the carrier. The idea of possible interference to this carrier signal is introduced, and the possible visible effects on the received signal.

History of television

Karl Braun invented the **cathode ray tube** (CRT) in 1897.

Boris Rosing suggested a system of transmitting pictures while he was working in St Petersburg, Russia, in 1906. The transmitter would have used mechanical scanning (with a special type of mirror drum) but the receiver would have used a CRT for the visual display. Rosing was granted a patent in 1907.

In 1912 Rosing and his assistant, Vladimir Zworykin, suspended work because they realised that the equipment they needed to progress was not available. At the start of World War I in 1914, Zworykin went to the USA. There he first joined the Westinghouse company, before moving on to RCA. In 1921 work was resumed on television, and in 1922 a US patent was granted for an all-electronic colour television system.

In the USA radio has always been run on a commercial basis. This is probably why Zworykin was prevented from starting a television service.

In the UK, Campbell-Swinton presented the fundamental ideas of an all-electronic system for the transmission and reception of pictures in 1924. His proposals were not realised until the early 1930s.

Low-definition pictures

John Logie Baird started his experiments on television in 1923. In 1925 at Hastings, on the English South Coast, he transmitted pictures between two of his machines. In 1926 he moved to London. Here he demonstrated to the Royal Institution images of moving human faces. These were not just plain black-and-white outlines, but had different shades of grey. The results were very crude, but it was just possible to make out individual faces.

Baird called his television equipment a **Televisor** (Figures 9.1–9.3). The mechanical device used to scan the frame was a **Nipkow disc**. This had been invented in 1884. Basically it was a rotating disc, in which were pierced a number of small holes on a spiral path. There was a photocell mounted behind the disc. As the disc rotated behind an aperture, 'scanning' of the scene took place.

Figure 9.1 A low-definition picture.

Figure 9.3 An early receiver.

The scene was scanned in a number of slightly curved vertical lines. Variations in the light falling on the photocell produced electrical variations. These variations were transmitted, received and reproduced by a suitable receiver.

The receiver therefore had also to contain a Nipkow disc, which when operating had to be synchronised with the disc in the transmitter. To overcome the problems of vibration, the mechanical system had to be quite rigid,

and this meant that the receiver usually had to be fixed to the floor.

The Baird company and the BBC started experimenting with a service of broadcasts in 1929. In August 1932 the BBC took over transmissions completely. The picture size was only 4 in × 2 in (100 mm × 50 mm). Each frame consisted of 30 lines, and the picture repetition rate was 12.5 per second. Picture detail was very poor, and the entertainment value was at a minimum.

Work on the Baird system continued. By 1936 better transmitting equipment had been developed. The improvements resulted in a picture with 240 lines per frame and a repetition rate of 25 per second. However, the scanning system was still mechanical.

Electronic scanning

Meanwhile, in 1931 the EMI research team led by Isaac Schoenberg had been working on the problems. Team members included A.D. Blumein and C.O. Brown. Another team member was J.D. McGee, recently made a fellow of the Royal Society. In 1936 this team, working with Marconi's Wireless Telegraph Company, had designed and produced an all-electronic system – that is, one with no moving parts. This system used 405 lines per frame. It scanned at 50 frames per second, **interlaced**. This produced 25 complete pictures every second. (The frequency of 50 per second frequency was chosen because the mains electricity supply in the UK is 50 Hz.

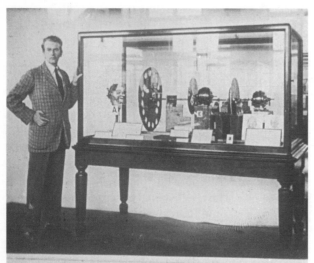

Figure 9.2 Baird and his Televisor.

This reduced vertical scan problems by locking the field timebase to the mains supply at the transmitter.)

In 1936 the BBC decided to give the two systems (mechanical and electronic) trial runs on alternate weeks. After about a year it was decided that the EMI all-electronic system should be adopted for public service. Baird's mechanical system was abandoned.

Baird was the innovator; he made the running, and obtained the first pictures. He aroused interest in others, but the television system used throughout the world today is not his.

The original theory was proposed by Campbell-Swinton. It was Zworykin who patented the electronic system, and Schoenberg's EMI team who created the television system we know today. Modern 405-line and 625-line services could never have been offered as a service using a system of mechanical scanning.

Converting scenes to electrical signals

A television camera system

When we look at a scene with our eyes, all the information in the scene is transmitted simultaneously to the brain. If a television system was designed using a similar principle the amount of bandwidth required would be far too great.

The main job of the television camera is to change the light reflected from the scene into electrical signals. The image in the camera is produced by scanning the scene in a way similar to the way our eyes read the pages of a book: from left to right and from top to bottom of the page.

Each small portion of the image is scanned separately to produce a voltage, which varies with the scene intensity: a higher voltage for a bright scene and a lower voltage for a dark scene. This varying voltage is called the **video signal**. It is used to **modulate** the main radio frequency carrier before transmission.

In effect the scene is broken down into a number of parallel lines, with each line producing a varying voltage at the camera output. In the television receiver's cathode ray tube (CRT) the received video signal is made to modulate the intensity of the electron beam while it moves across the screen (left to right) and down (top to bottom) at the same time, building up the picture line by line in what is called a **raster**.

Obviously, in the camera the image must be read again and again at a very high rate so that a continuous picture appears on the CRT. To produce a picture of acceptable quality at least 500 scanning lines are required. Also, to reduce the effect of 'flicker', the picture must be repeated at least 50 times a second.

At the cinema, pictures are projected on the screen at 24 frames per second, but each frame is seen twice thanks to a segmented disc – shaped rather like a gentleman's bow-tie – which rotates in front of the projection lens to increase the picture or flicker rate to 48.

To reduce the effect of flicker, a television system relies on the help of the **persistence of vision**. The eye retains the impression of an image on the retina for a very short time after the light from the image has ceased. The result is that the separate frames appear to merge to create the impression of a continuous picture.

The studio TV camera

Figure 9.4 shows a typical device used for converting light from a scene into an electrical signal. The **vidicon** camera tube has been used in broadcasting since the 1950s for studio and closed-circuit applications. There are other types of camera tube, but the small size of the vidicon tube (about 25 mm in diameter and 175 mm in length) has made it very popular, especially for use in early colour cameras.

The image produced through the lens system is focused onto the **photoconductive layer** or **target**, which is on the right-hand side of the transparent signal plate in Figure 9.5b.

A photoconductive material is one that conducts electricity when light falls upon it. If there is a voltage across such a material, a small current will flow even in complete darkness. This is the **dark current**. It should be as small as possible to give a good black level. Conduction increases as the level of illumination is increased, but the relationship between the illumination and the increase in current is not linear. This non-linearity will be mentioned again in connection with **gamma correction**.

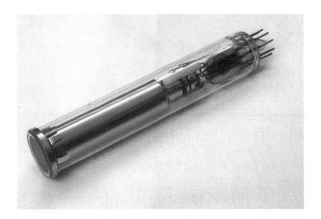

Figure 9.4 Vidicon tube as used in a camera.

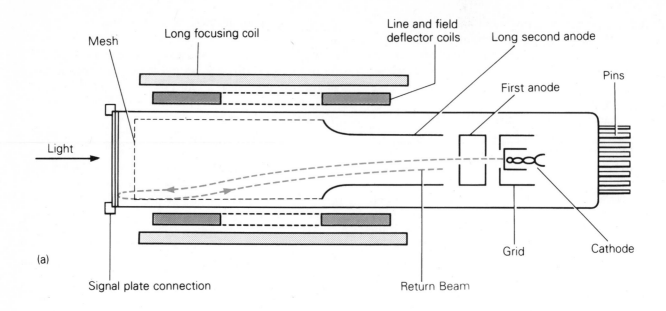

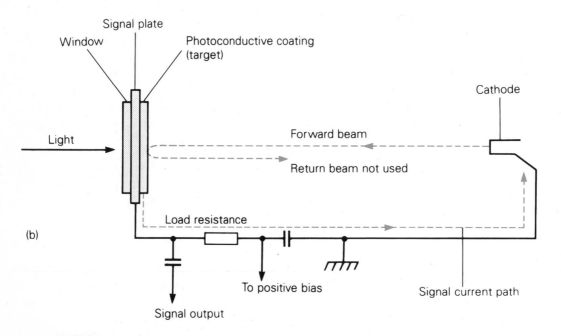

Figure 9.5 (a) Cross-section of vidicon tube, and (b) equivalent circuit.

The vidicon uses antimony trisulphide as the photoconductive coating. The target assembly is very small – about 400 mm^2 in area. Because the signal plate (which is conductive) has to be optically transparent, it is extremely thin, as is the photoconductive coating (about 10 μm).

The photoconductive target is scanned by an electron beam emitted from the cathode. As in all thermionic camera tubes and cathode ray tubes the beam must be

focused and also deflected. Magnetic deflection and focusing are used for this type of tube. The deflection coils are beneath the focusing coil, which is a long solenoid that runs almost the entire length of the tube. The electron gun is designed so that the beam strikes the target at right angles. This is known as **orthogonal scanning**.

The target is at cathode potential, but the signal plate is biased with a positive voltage of say 20 V. Capacitance exists between the gun side and the signal plate side of the photoconductive layer, so that with no light from the camera lens, a charge of 20 V appears across it from the target bias source. When the illuminated scene is focused on the target, a brightness pattern is formed on the signal plate side, and electrons migrate from the gun side to the signal plate side of the photoconductive layer.

More electrons move to the brighter parts than to the darker parts. This means that on the gun side of the layer there is a shortage of electrons, corresponding to the brightness of the optical image on the signal plate side: the brighter the area, the greater the shortage of electrons, and the higher the positive potential on the gun side. Therefore on the gun side of the photoconductive target there is an electrical replica of the optical image, in the form of more or less positively charged areas.

It takes time for any electron migration to take place, depending on the capacitance and resistance of the target material. As any point is scanned by the beam once per picture period, any change in the screen brightness should be able to show itself as an change of electrical image within that time.

The areas of the electrical image that are most positive – that is, short of electrons – take electrons from the beam to neutralise their charge. The quantity of electrons taken depends on the light on the optical side of the target. The beam current must vary accordingly. This variation takes place in the path external to the tube, as shown in Figure 9.5b. Voltage changes are set up across the load resistor, to provide an electrical representation or 'voltage copy' of the studio scene.

Electrons taken up by the target flow in the external circuit. Those not taken up return to the first anode. This return beam must vary in density in the same way as the external beam current, as the two together add up to form the entire beam current. The modulation on the return beam will, however, be in the opposite sense to that through the external load, because taking away electrons for bright areas leaves fewer electrons in the return beam, and vice versa. The modulated return beam is not used in this type of camera tube, but it is used in a type known as an **image orthicon**.

Typical voltages for the electrodes for the vidicon tube are:

Grid, for beam intensity: variable – 40 V
First anode: 300 V
Second anode, variable, for beam focus: 250 V

The signal plate provides the target bias control. This functions as an amplitude control, as the potential between the plates of the target material affects the electron migration. It is biased positively and is also variable.

1 Inspect different types of vidicon tube. Identify the electrodes within the glass envelope: for example, the second anode, the signal plate connection, the grid and first anode assembly.

2 Identify components outside the tube, such as the focusing coil and the line and field deflection coils.

3 Connect an oscilloscope to the video output socket on the camera. Set the oscilloscope timebase to 10 ms/division. Synchronise the oscilloscope to show one line of video signal. Now move the camera position and note how the video signal on the oscilloscope changes as the camera moves.

Dividing up a scene: scanning

As we have already explained, the **raster** is the scanning pattern of the camera target or the cathode ray tube screen. It is caused by an electron beam striking the target or the tube screen phosphor while being deflected horizontally and vertically. In the receiver the raster is present whether or not there is a video signal. If you remove the video signal from the TV receiver or monitor and advance (turn up) the brightness control, you will be able to see the raster.

Simple scanning

In **simple** or **sequential scanning** of the CRT (Figure 9.6) the electron beam or spot moves across the screen, tracing out the first line, A–B. The force produced by the field

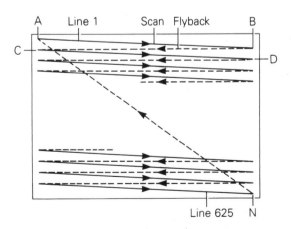

Figure 9.6 Simple (sequential) scanning.

timebase makes the scan lines slope down: as the beam is moving across the screen it is also being pushed down. At point B, it returns to the left-hand side at a much faster rate (**line flyback**) to point C below the start of line A–B.

Line 2 starts at point C and continues to point D. This process continues until the beam reaches point N. The complete movement of the beam, starting at the top and finishing at the bottom, is known as a **field scan**. At point N, the beam moves back to point A at the top of the screen (**field flyback**). The beam now recommences a new field scan to produce the second field, and the process repeats.

In the method just described, 625 lines would be scanned. This process would be repeated 25 times a second, producing a field rate or **picture rate** of 25 Hz. The spot of light moves so fast that the eye sees a continuous picture. The 'after glow' of the CRT phosphor and the eye's persistence of vision assist this effect.

Interlaced scanning

No scanning system is perfect. Even at a picture rate of 25 Hz the eye would notice an objectionable 'flicker'. One way to solve this problem would be to increase the picture rate or field rate. As we shall see later, the overall bandwidth of the television picture channel is proportional to the number of pictures radiated per second. So if we double the number of pictures per second then we also double the radiated bandwidth.

However, it is possible to increase the flicker rate to 50 Hz while keeping the picture rate at 25 Hz and not increase the bandwidth at all. This is achieved by using a modified form of scanning called **interlaced scanning**.

In interlaced scanning, the beam moves as before, but this time leaves a greater gap between the lines traced. In the first field only half the 625 lines are scanned: that is, 312.5 lines.

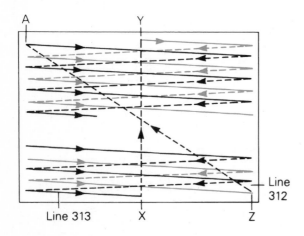

Figure 9.7 Idealised interlaced raster.

Figure 9.7 shows an idealised interlaced raster. After completion of the first field (shown in black which consists of 312.5 lines) at point X, the field timebase causes the beam to return to the top centre of the screen (point Y). For the next field scan (shown in red in Figure 9.7) the second half of line number 313 begins at point Y. From this position, all successive lines are made to fill in the gaps that were left by the first field. This second field finishes at point Z. Once again field flyback occurs, causing the beam to return to the starting point A.

There are two sets of forces at work here. The **line timebase** causes horizontal movement of the beam (scan and flyback); the **field timebase** causes vertical movement (scan and flyback).

The combined forces of the two timebases produce two interlaced fields on the CRT screen, each consisting of 312.5 lines. It takes 1/50th of a second to complete each field. Each complete picture requires two of these fields, and each picture is repeated 25 times a second.

As the picture repetition rate is 25, the flicker rate is equal to $2 \times 25 = 50$. Therefore the field timebase runs at a frequency of 50 Hz, and is responsible for 50 downward strokes of the spot per second. Each downward stroke of the spot produces 312.5 lines and therefore $312.5 \times 50 = 15\,625$ scanning lines per second. This is also the operating speed of the line timebase. The time taken for 1 line scan and flyback is equal to $1/15\,625$ seconds = 64 μs.

Interlaced scanning reduces the amount of flicker on the screen because the total area of the screen is covered at twice the rate: 50 times a second rather than 25 times a second. If you examine the screen of a TV receiver closely, you will notice that individual lines still suffer from a 25 Hz flicker effect, but at normal viewing distances the overall effect is quite tolerable, because doubling the vertical movement of the spot from 25 to 50 times a second in the interlaced scanning system gives the effect of 50 pictures a second.

Lost lines

Figure 9.7 shows the path taken by the beam during field flyback. As we have already stated, this is an idealised raster pattern. It assumes that the beam moves from the bottom of the screen to the top instantaneously. In practice, while the beam is travelling upwards it is also deflected from left to right by the line timebase. As a result, lines are 'lost', and unable to carry picture information. The actual number of lines carrying picture information is reduced from 625 per picture to 575. Therefore 50 lines are lost per picture, or 25 lines per field. During field flyback, the beam takes approximately

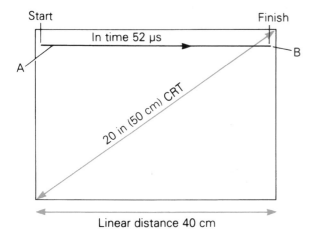

Figure 9.8 Calculation of spot velocity.

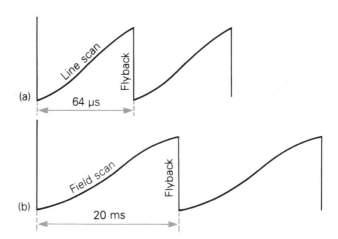

Figure 9.9 Requisite waveforms. (a) Line scan; $\tau =$ 64×10^{-6} s, frequency = $1/\tau$ = 15 625 Hz. (b) Field scan; $\tau = 20$ ms = 20×10^{-3} s, frequency = $1/\tau$ = 50 Hz.

1.5 ms, which is well within the time taken for the duration of 25 picture lines: 25×64 ms = 1.6 μs.

Spot velocity

Figure 9.8 shows the front of a 20 in (50 cm) CRT. For line scan it takes 52 μs (64 μs – 12 μs flyback) for the beam to travel from A to B. The linear distance for this size of tube is approximately 40 cm. Therefore:

$$\text{Speed, in m/s} = \frac{\text{Distance travelled}}{\text{Time taken}}$$
$$= \frac{40\,\text{cm}}{52\,\mu\text{s}}$$
$$= \frac{40 \times 10^{-2}\,\text{metres}}{52 \times 10^{-6}\,\text{seconds}}$$
$$= 7692\,\text{m/s}$$

This is equivalent to a speed of more than 17 000 miles per hour for a CRT of size 20 in (50 cm). Beam flyback to the left-hand side of the tube is even faster. The larger the cathode ray tube, the faster the spot has to travel in the same amount of time.

Sawtooth waveforms

The two timebases (line and field) must each be capable of producing suitable current waveforms that will give linear beam movement on the screen, left to right and top to bottom. Figure 9.9 shows the requisite shapes: as you can see, they are basically 'sawtooth'.

The composite video signal

In any broadcast or closed-circuit television system, the two electron beams (the one in the camera and the one in the receiver CRT) must be **time coincident**: that is, they must be in the same position at the same time. When the camera is scanning the beginning of line 1, then so must the receiver CRT. If it is not, then the video signal will appear on the screen in the wrong position.

The camera timebases have to **synchronise** the receiver timebases. This is done by sending out timing pulses with the video signal. These timing pulses are added to the video signal either inside the camera or at some later stage prior to transmission. They are known as **line-** and **field-synchronising pulses**.

Figure 9.10 shows one line of video signal complete with two line sync pulses. The time reference point is the leading edge of the line sync pulse. The complete line period (sometimes given the letter H) is 64 μs. Approximately 12 μs is **blanked out** – that is, it contains no picture information – to allow for the insertion of the line sync pulse.

The actual line sync pulse is preceded by a front porch and followed by a back porch. The **front porch** allows time for any voltage due to picture content to fall to blanking level before the line sync pulse commences. This ensures good sync. It is especially important when the picture content at the end of a line is white. The line-synchronising pulse will initiate line flyback in the receiver almost immediately, causing the CRT beam to deflect from right to left. This flyback takes about 10 μs, so the picture information must be kept blanked until the start of the forward scan for the next picture line.

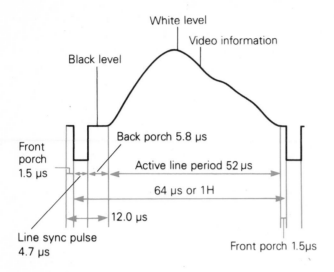

Figure 9.10 Composite video signal waveform for the UK 625-line system.

The **back porch** provides additional blanking. It is included to allow time for the receiver line flyback to be completed before picture information is introduced for the next line.

Time intervals may be taken as follows:

Front porch: 1.5 μs
Back porch: 5.8 μs
Line sync pulse: 4.7 μs
Hence line blanking: 12.00 μs

The **active time** of a complete line is the part that contributes towards picture information: that is, 64 μs minus the time taken for the line sync pulse and the front and back porches (total 12 μs), or 52 μs.

See Figure 9.11. Field flyback must be initiated at the end of each field to return the beam from the bottom to the top of the screen. Twenty-five lines are set aside in every field for this to happen. With **field blanking** of 25 lines, there are 312.5 − 25 = 287.5 active lines per field. The **pulse trains** inserted at the end of every field are complicated, because of the use of interlaced scanning and the need to maintain line sync. The main features are as follows:

1. first set of **equalising pulses**, which allow time for the voltages in the field sync circuitry to be at the same level at the end of each field – that is, before the field sync pulses occur;
2. second set of equalising pulses, which allow time for

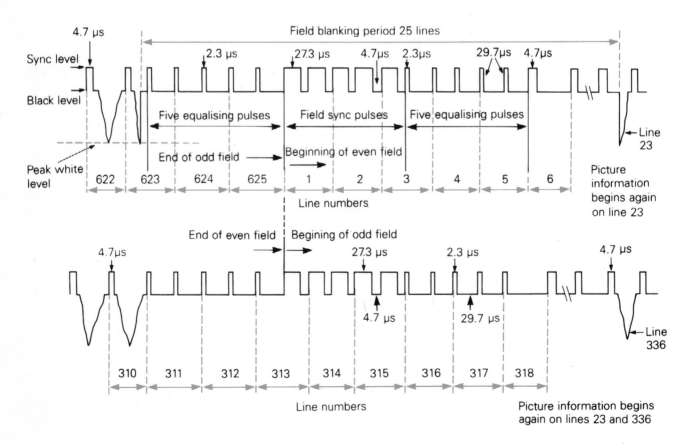

Figure 9.11 Field-synchronising waveforms of the UK 625-line system.

the sync circuit voltages to be the same in each field, after the set of field sync pulses;

3. **field sync pulses**, which are integrated to form a single sync pulse that initiates the field flyback;

4. **blanked lines**, approximately 17, which are blanked to allow completion of the field flyback before the resumption of picture information.

Many of the world's broadcasters insert data onto the 17 blanked lines mentioned above. Anyone can use this data as long as they have a suitable receiver with a built-in decoder. In the UK this data service is known as **teletext**. Data is sent out with all the regular programmes, and it is designed so that it does not interfere with them. Insertion test signals are also contained on some of these lines.

ACTIVITIES

4 Set up a black-and-white television receiver to display a still picture. A test pattern from a generator will do, or an off-air test card from your aerial system. Reduce the picture height by about 3 cm using the picture height control. Now inspect the top or bottom of the picture closely. You should see that flicker is more noticeable in these areas. If you now move away from the screen you should notice that the effect of the flicker reduces, making the picture more tolerable to watch.

5 Determine the velocity of the electron beam travelling across the inside face of a CRT. Try this on a 12 in (30 cm) tube and a 26 in (66 cm) tube. Use the formula given earlier for spot velocity.

6 If a current probe is available you can connect it to the deflection coils (or their connecting wires) of a black-and-white TV receiver. The coils are mounted on the tube neck.

 Beware: this is a high voltage area!

Connect the other end of the current probe to the Y channel of an oscilloscope. If the probe is connected to the line deflection coils, then the oscilloscope's timebase should be set to about 10 μs/division. If it is connected to the field deflection coils, then the timebase should be set to about 5 ms/division. If all goes well, you should see a sawtooth-shaped voltage waveform (look back at Figure 9.9).

The voltage waveform is in fact a copy of the current within the conductors. The current probe works by picking up the magnetic field that surrounds the conductor: from this magnetic field a voltage is induced inside the pickup coil. This very small voltage is then fed to an amplifier inside the probe and then on to the Y input on the oscilloscope.

4 Connect an oscilloscope to the output of a video camera and show one line of video signal. Using the timebase settings on the oscilloscope, try and work out the time duration of the following:
   i) the line sync pulse;
   ii) the front porch;
   iii) the back porch;
   iv) the active part of the complete line, video only.

All your answers should be in microseconds.

Video bandwidth

Look at the screen patterns shown in Figure 9.12. Notice how the rate of change of video signal (frequency) increases as the pattern becomes finer. In order to reproduce a very fine pattern then a high video frequency is required (Figure 9.12c). Coarse patterns, such as the one shown in Figure 9.12a, require low video frequencies.

Bear in mind that the highest video frequency will have a large say in determining the overall bandwidth of the complete television channel: that is, the vision and sound carriers, with their respective bandwidths.

Spot size

It would be wasteful of bandwidth to transmit a higher-frequency video signal that might not be capable of being seen owing to the limitations of the size of the CRT spot.

As the highest video frequency dominates this discussion on bandwidth, we must consider a theoretical screen pattern composed of a large number of alternate black and white squares (a **chequer-board pattern**), as shown in Figure 9.13. The greater the number of squares the better, as the area of each square places a limitation on the maximum vertical and horizontal resolution (fineness of detail)

The vertical resolution is determined by the number of lines in the particular TV system. The horizontal resolution is determined by the highest video frequency.

Let us assume that our TV screen is scanned by 625 lines. In the 625-line television system 50 lines are not

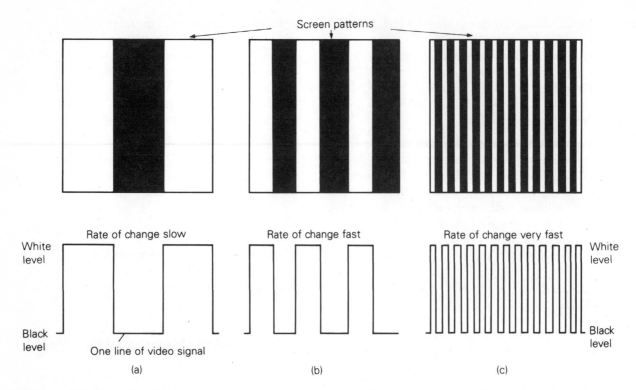

Figure 9.12 Video signal drive waveforms required to reproduce the patterns shown; one line of video signal shown in each case.

used for picture information. Therefore our calculation will be based on a figure of 575 lines. The maximum number of vertical picture points will be $625 - 50 = 575$. For a square TV screen an equal number of picture points would be obtained in both directions. However, a TV screen is not square but rectangular, and has an **aspect ratio** of 4 : 3 (four units by three units), as shown in Figure 9.14.

The aspect ratio is the ratio of the width of the television screen to the height of the television screen. The agreed standard for this is a ratio of 4 : 3.

The total number of picture points or elements in one picture $= 575 \times 575 \times 4/3 = 440\,833$ elements. There are 25 complete pictures every second, so this becomes $440\,833 \times 25 = 11\,020\,825$ elements in one second.

Figure 9.15 shows a small section of the television screen; one picture line is shown. To reproduce just one black/white section requires one cycle of video signal

Figure 9.13 Chequer-board pattern; square screen.

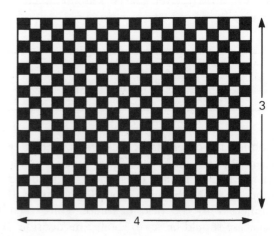

Figure 9.14 Chequer-board pattern; aspect ratio of 4 : 3.

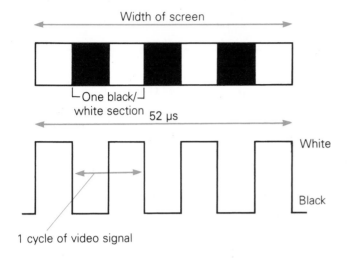

Figure 9.15 Video signal for one picture line.

waveform. To complete the calculation, the highest video frequency is now equal to $11\,020\,825/2 = 5\,510\,412\,\text{Hz} = 5.5\,\text{MHz}$. Therefore the total video bandwidth becomes 0 Hz to 5.5 MHz.

As mentioned earlier when discussing scanning, using 50 pictures per second with each one made up of 575 lines would in effect double the system bandwidth from 5.5 MHz to about 11 MHz.

So you can now see that using interlaced scanning halves the system bandwidth while at the same time increasing the flicker rate to 50.

ACTIVITIES

7   Connect an oscilloscope to the video output socket of a TV pattern generator. Show one line of video signal on the oscilloscope. Set the timebase at 10 μs/division. Alter the pattern from coarse to fine and note how the video signal waveform also becomes finer. Suggested patterns for this activity are a chequer-board pattern as in Figure 9.14, or a crosshatch pattern (a fine white grid on a black background).

8   Check the aspect ratio of a domestic television receiver of your choice as follows. Measure the viewable width of the screen – where the picture begins and ends (left to right) – and the viewable height of the screen – where the picture begins at the top and finishes at the bottom. Don't include the outer thickness of glass in your measurements. Use the same units for both measurements Now work out the aspect ratio by dividing the width measurement by the height measurement. Your result should be a figure of about 1.3. This represents an aspect ratio of about 4 : 3. In other words, the width is about 1.3 times the height.

CHECK YOUR UNDERSTANDING

- A television camera is a device for converting light from a scene into an electrical signal. There are many types. The type using the vidicon tube is one of them.
- Orthogonal scanning is used within the vidicon tube.
- To produce a picture of adequate quality, a minimum of 500 scanning lines are required. To reduce the flicker effect, a television system relies on the help of the persistence of vision.
- The resistance of a photoconductive material decreases as the light falling on it increases.
- The raster is the scanning pattern of lines produced by an electron beam.
- In simple scanning, 625 lines would be scanned all at once.
- One complete set of horizontal lines starting at the top of the screen and ending at the bottom is known as one field scan.
- The line timebase causes horizontal movement of the beam and the field timebase causes vertical movement. The combined forces produce a raster.
- Interlaced scanning is used to increase the flicker rate to 50, while keeping the picture rate at 25, without increasing the system bandwidth. For every picture there are two fields, and each picture is repeated 25 times a second.
- In the 625-line system, 50 lines are not used for picture information. This leaves a total of 575 lines.
- In the receiver, sawtooth-shaped currents produced by the two timebases give linear movement of the beam on the screen. One single line scan and flyback takes 64 μs. One single field scan and flyback takes 20 ms.
- Line and field sync pulses are transmitted with the video signal so that the transmitting and receiving timebases can keep in step with each other. Line sync pulses are sent out after every picture line, and field sync pulses are sent out after every field.
- Line sync pulses last for 4.7 μs. Field sync pulses last for 160 μs and are much more complex, owing to interlaced scanning.
- The active part of a complete line – that is, the part of the complete line that contains the video signal – is 52 μs, 12 μs being allowed for the front porch, back porch and line sync pulse.
- Coarse picture patterns require low video frequencies; fine picture patterns require high video frequencies. The higher the video signal frequency used, the greater the system bandwidth required.
- The horizontal resolution is determined by the highest video frequency, and the vertical resolution is determined by the number of lines used in any particular system.
- The aspect ratio is the ratio of the width of the screen to the height of the screen. In all television systems, an

aspect ratio of 4 : 3 is used.
- For system I, 5.5 MHz represents the highest video frequency that could be transmitted.

REVISION EXERCISES AND QUESTIONS

1 What is the raster flicker rate for the British 625 line system I?

2 Explain how the flicker rate is increased from 25 to 50 times a second.

3 In the 625-line TV system, why are there a number of 'lost lines'?

4 What is the purpose of the synchronising pulses that are transmitted with the video signal?

5 Describe the meaning of the term 'persistence of vision'.

6 What is the aspect ratio of the television receiver screen?

Transmitting the video signal

Introduction

This chapter deals with the way the video signal is actually transmitted. We look at the ways in which certain problems are overcome. We also compare the television standards used by various countries throughout the world.

AM transmission

All terrestrial television transmissions use **amplitude modulation** for the video signal. If the radio-frequency carrier is modulated with a single modulating signal, the basic carrier and two side frequencies are generated. Modulating signals are normally complex, consisting of a range of basic waveforms. Bands of frequencies therefore exist on either side of the carrier: the **upper sideband** and the **lower sideband** (Figure 10.1).

The disadvantage of this method of modulation is that it doubles the amount of space taken up in the available spectrum. For AM radio (UK and Europe medium waveband) this does not matter too much, because each radio channel requires only about 9 kHz (+4.5 kHz).

The medium waveband begins at approximately 500 kHz and finishes at approximately 1600 kHz. So the amount of spectrum used is about 1.1 MHz (1600 kHz − 500 kHz). If we now divide the available spectrum (in this case 1 MHz) by the amount of space taken by one medium-wave channel (9 kHz), we get some idea of the number of channels that we can fit in. So, 1.1 MHz ÷ 9 kHz = 122. As you can see, 122 channels can be fitted into the available spectrum of 1 MHz. In reality, the number is much less than this.

For video signals, the existing frequencies extend from d.c. (that is, 0 Hz) up to 5.5 MHz (UK system I). If the radio-frequency carrier is modulated with a complex signal that has frequency components up to 5.5 MHz, then the carrier and many sidebands will be generated. These sidebands will finish at 5.5 MHz above and 5.5 MHz below the carrier (Figure 10.2). The total space

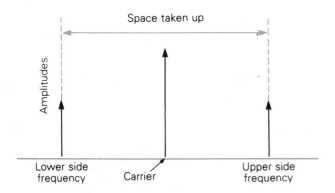

Figure 10.1 Lower and upper side frequencies.

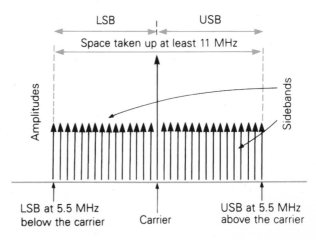

Figure 10.2 The use of normal double-sideband transmission.

taken up by the channel is 11 MHz. This type of transmission is known as **double sideband transmission**.

The space occupied by this single vision channel would need to be greater than that shown (11 MHz). This is due to the shape of the slope produced by the filters, whose job it is to prevent modulating signals above 5.5 MHz from generating any unwanted sidebands. About 0.5 MHz is actually wasted on either side of each sideband. Also, each single television channel must have its own sound carrier, which must be positioned outside the extremities of one of these sidebands. This can add a further 0.25 MHz. Together with the 1 MHz allowed for the sidebands, this gives a total of 1.25 MHz. So the original figure of 11 MHz is now increased to 12.25 MHz.

Vestigial sideband transmission

With amplitude modulation, both sidebands contain the same information. Therefore, as long as the carrier is maintained, one of the sidebands may be removed altogether. The result would be a much reduced transmission channel, comprising the carrier and one of the sidebands (Figure 10.3).

As we have already stated, video signal frequencies extend from 0 Hz or d.c. right up to 5.5 MHz. It would be very difficult to design a filter capable of removing one complete sideband without affecting the carrier and all the low-frequency signal components. Therefore a compromise approach is adopted, in which most of one of the sidebands is removed before transmission. What is actually transmitted is one complete sideband and a small amount of the other sideband. This type of transmission is known as **vestigial sideband transmission** (the word 'vestige' means a remainder or trace).

Figure 10.4 shows the transmission spectrum of the UK 625-line TV system I. For comparison, Figure 10.5 shows a double-sideband transmission using the same standards.

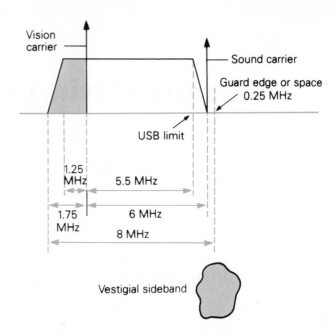

Figure 10.4 The use of vestigial sideband transmission: a saving of 4.25 MHz.

As you can see, about 8 MHz is allocated for a single television channel using vestigial sideband transmission. This is a big saving in bandwidth compared with the original figure of 12.25 MHz.

Frequency modulation is used for the sound carrier.

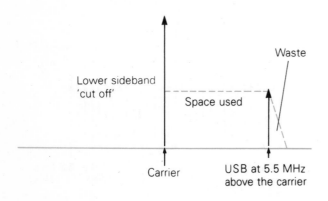

Figure 10.3 The use of single-sideband transmission.

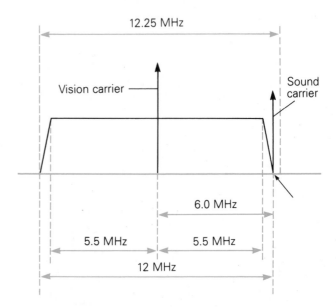

Figure 10.5 RF spectrum, showing the effect of using double-sideband transmission, compared with vestigial sideband.

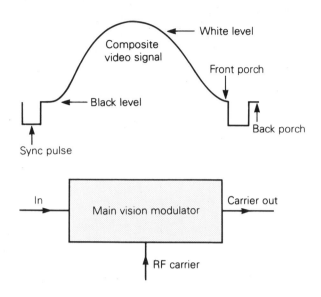

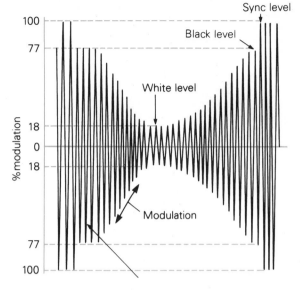

Figure 10.6 Negative vision modulation.

The deviation is ±50 kHz, and the distance between the sound carrier and the vision carrier is 6 MHz (more accurately 5.996 MHz). The bandwidth required for the sound carrier is 200 kHz.

Negative and positive vision modulation

Figure 10.6 shows the RF envelope at the output of the vision modulator in the TV transmitter. Peak white

corresponds to about 18 per cent of the carrier value, and the sync pulse tip corresponds to 100 per cent of the carrier value. Black level is at a value of 77 per cent. Therefore decreasing values of picture brightness mean that the carrier amplitude is increased in value. Pictures with bright scenes decrease the value of the carrier to a minimum of 18 per cent. This kind of modulation is known as **negative vision modulation**.

Figure 10.7 again shows the RF envelope at the output of the vision modulator, but the video information is reversed in polarity. Peak white or maximum brightness now corresponds to maximum or 100 per cent carrier

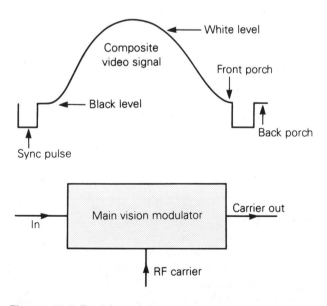

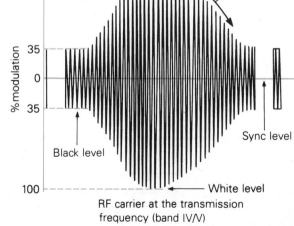

Figure 10.7 Positive vision modulation.

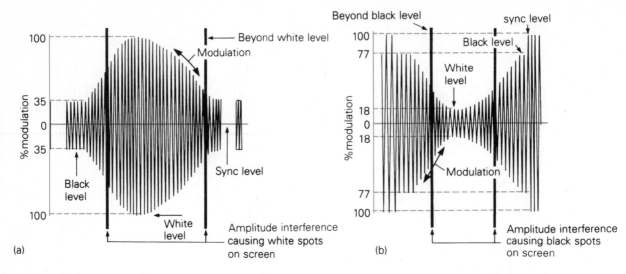

Figure 10.8 Amplitude interference: (a) positive vision modulation; (b) negative vision modulation.

amplitude. Lower degrees of brightness correspond to proportionately lower carrier amplitudes. At black level the carrier is at an amplitude of 35 per cent. The bottom of the sync pulses takes the carrier down to 0 per cent. This kind of modulation is known as **positive vision modulation**.

In the now obsolete UK 405-line television system, positive vision modulation was used. It was found to be unsatisfactory because of the presence of **amplitude interference**. This can cause erratic and sudden changes in the amplitude of the RF carrier. As a result, random white spots appear on the receiver screen (Figure 10.8a). When negative vision modulation is used, amplitude interference would cause black spots to appear on the receiver screen (Figure 10.8b).

It is less objectionable to the human eye to have black spots than white spots on the receiver screen. This is because the persistence of vision means that a white spot is 'seen' on the screen for a long time after the interference has finished. This is why negative vision modulation is now almost universally used.

The only disadvantage of using negative vision modulation is that the problems caused by amplitude interference on the synchronising pulses could cause poor triggering of the receiver line timebase. As we shall see in later chapters, when we look at the receiver circuitry, the above problem can be overcome by using a special 'flywheel' sync circuit. This has the advantage of not relying on every line sync pulse, but rather on a succession of pulses. If interference causes any of the line sync pulses to be 'lost', this does not then cause any significant problem. Now that most countries are using UHF rather than VHF for the RF carrier, then as long as the signal strength is adequate, interference becomes less of a problem.

Basic transmission arrangement

Figure 10.9 shows the block diagram of a basic vision transmitter. It is in two sections: one deals with the generation of the composite video signal (the camera and the sync pulse generator), and the other deals with the generation of the RF carrier, complete with RF power amplifier and vestigial sideband filter. The complete diagram is of course a simplified version of the real thing, but it does show how the signals are developed from the camera to the aerial.

World television standards

Table 10.1 lists the different types of television system in use worldwide, and Table 10.2 lists the television channels and nominal carrier frequencies in the UK. Note that channels 35, 36, 37 and 38 are not used by broadcasters at present, as they are reserved for home entertainment goods such as videocassette recorders and computer games. These give out radio frequency signals, which can be adjusted within the range of channels 35 and 38.

However, there are now discussions going on that may change this. Because of the lack of space available in the bands, permission may be given to use two of these spare channel positions. This would provide the UK with a fifth **analogue** television channel (channel 37) and four terrestrial **digital** television channels. The digital channels would fit into the existing 8 MHz allowed for each analogue channel (in this case channel 35). If this causes interference with domestic

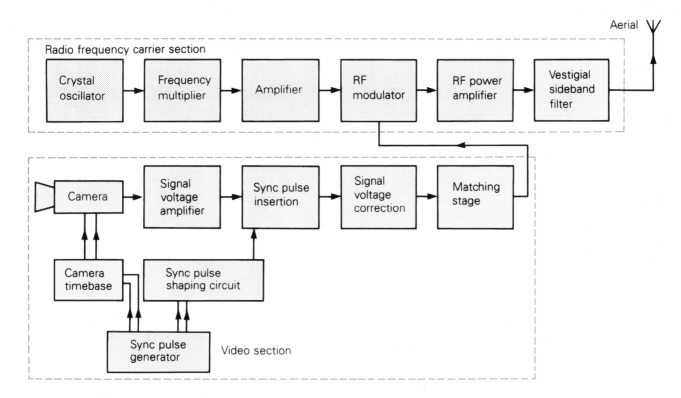

Figure 10.9 Simplified vision transmitter.

equipment, funds would be available to pay for teams of technicians to visit all the relevant sites in the UK to retune video recorders. The view of the broadcasting authorities is that there may be problems of adjacent channel interference. At the time of writing, a final decision had not been made.

Figure 10.10a shows two adjacent UHF channels, 61 and 62. Note that they are very close together. This fact emphasises the need to conserve as much bandwidth as possible. Channels 61 and 62, or any two adjacent channels, would never be broadcast from the same transmitter because they would cause interference within the receiver.

In 1961 it was agreed under the **Stockholm Frequency Plan** that each transmitter would be allocated four television channels. These would fall within a frequency range of 88 MHz. In theory, for a frequency range of 88 MHz, 11 television channels can be accommodated

Table 10.1 World television standards

System	Lines	Channel width (MHz)	Bandwidth (vision) (MHz)	Separation (MHz)	Vestigial sideband (MHz)	Vision modulation	Sound modulation
B	625	7	5	+5.5	0.75	Negative	FM
D	625	8	6	+6.5	0.75	Negative	FM
G	625	8	5	+5.5	0.75	Negative	FM
H	625	8	5	+5.5	1.25	Negative	FM
I	625	8	5.5	+5.996	1.25	Negative	FM
K	625	8	6	+6.5	0.75	Negative	FM
L	625	8	6	+6.5	1.25	Positive	AM
M	525	6	4.2	+4.5	0.75	Negative	FM
N	625	6	4.2	+4.5	0.75	Negative	FM

Note: For Systems L2, L3 and L4 vision/sound separation is –6.5 MHz (France) (Reproduced with permission from the 1995 edition of *The World Radio and TV Handbook*, Billboard Books.)

Table 10.2 Television channels and nominal carrier frequencies for the UK

Channel numbers	Carrier frequencies (MHz)	
	Vision	Sound
Band IV UHF		
21	471.25	477.25
22	479.25	485.25
23	487.25	493.25
24	495.25	501.25
25	503.25	509.25
26	511.25	517.25
27	519.25	525.25
28	527.25	533.25
29	535.25	541.25
30	543.25	549.25
31	551.25	557.25
32	559.25	565.25
33	567.25	573.25
34	575.25	581.25
Band V UHF		
39	615.25	621.25
40	623.25	629.25
41	631.25	637.25
42	639.25	645.25
43	647.25	653.25
44	655.25	661.25
45	663.25	669.25
46	671.25	677.25
47	679.25	685.25
48	687.25	693.25
49	695.25	701.25
50	703.25	709.25
51	711.25	717.25
52	719.25	725.25
53	727.25	733.25
54	735.25	741.25
55	743.25	749.25
56	751.25	757.25
57	759.25	765.25
58	767.25	773.25
59	775.25	781.25
60	783.25	789.25
61	791.25	797.25
62	799.25	805.25
63	807.25	813.25
64	815.25	821.25
65	823.25	829.25
66	831.25	837.25
67	839.25	845.25
68	847.25	853.25

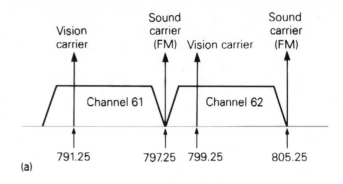

(a)

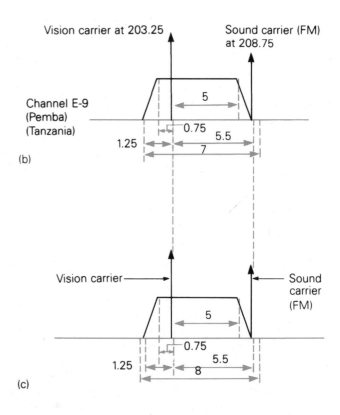

(b)

(c)

Figure 10.10 Systems B and G ((b) and (c) respectively) are similar to each other, but system G is used on UHF; therefore a wider channel space is allowed.

(88 ÷ 8 = 11). Using only four of these 11 channels means that they can be 'spaced' further apart from each other, to prevent the effect of adjacent channel interference.

For example, the IBA Winter Hill transmitter in the northwest of England serves the Greater Manchester area, plus Liverpool and the Wirral. The channel numbers used by this transmitter are 55, 59, 62 and 65. The respective frequency ranges are:

Channel 55, BBC1: 742–750 MHz
Channel 59, ITV: 774–782 MHz
Channel 62, BBC2: 798–806 MHz
Channel 65, C4: 822–830 MHz

Table 10.1 also shows that there are eight different versions of the 625-line system. No one 625-line television system is fully compatible with another. For example, system B uses negative vision modulation and FM for the sound carrier, while system L uses positive vision modulation and AM for the sound carrier. Systems K and L differ similarly. There are subtle differences between systems B and G/H, where the vestigial sideband limits have different cut-off frequencies. However, all systems have some common features, such as the aspect ratio and the use of interlaced scanning.

Different methods are used to transmit colour television signals. There are basically three types of system in use: the American **NTSC** (National Television Systems Committee), the German **PAL** (phase alternation line), and the French **SECAM** (séquentiel couleur à mémoire).

The NTSC system is used throughout the Americas, Japan and the West Indies. SECAM is used in France, Luxembourg, Russia and China. PAL is used in the UK, Germany, Africa, South Africa, Australia and many more countries. Of all the systems shown in Table 10.1, systems I and G/H are probably the closest in relation to each other.

A receiver designed for the PAL system I will function to a certain extent when operating from a PAL transmission system G/H (provided that the RF carrier signal used falls within the tuning range of the receiver), but there will be no sound. This is because the intercarrier sound signal will be at a different frequency, 5.5 MHz instead of 6 MHz. We shall explain intercarrier sound later, when we look more in detail at the receiver block diagram.

Figure 10.10 shows a comparison of the transmission spectrums of systems I, B and G.

Appendix 1, at the back of this book, lists most of the world's countries, together with their television systems. Some remote areas, such as those in the South Pacific, are still without their own television service. These areas usually rely on satellite systems for a service.

ACTIVITIES

1 For this activity you will need a spectrum analyser capable of receiving signals in the VHF and UHF bands, plus a strong signal from an aerial system. Connect the spectrum analyser to the aerial system with the correct leads. When the analyser is correctly adjusted and tuned in to your frequency band, you

should see two vertical spikes on the screen. These spikes indicate the presence of two RF carriers, one the vision carrier and the other the sound carrier. They show that RF energy is present at these frequencies. If the analyser is correctly calibrated you should be able to estimate the frequency of these carriers.

A spectrum analyser is basically a radio receiver combined with an oscilloscope (see Figure 10.11).

2 This activity will probably require assistance from your tutor. It is designed to show the shape of a negatively modulated vision signal. For this you will need a monochrome TV receiver, an oscilloscope (minimum bandwidth 40 MHz) and a good aerial system. Tune in the TV for normal picture and sound. Connect the oscilloscope using a high-impedance probe to the last IF stage, but just before the vision demodulator. A good test point would be the collector of the last IF stage or the input to the vision demodulator IC.

Set the timebase to about 20 μs and then synchronise the oscilloscope to show the IF envelope. The IF signal should be at a frequency of approximately 38 MHz–40 MHz (39.5 MHz). If you have any problems in getting a stable trace on the oscilloscope, then you could trigger it by using the unused Y input channel. (This assumes of course that you are using a double-trace oscilloscope). Don't forget to set the trigger selector to this second channel. Connect the second probe lead to the input of the video output stage, and you should now obtain a stable trace on the oscilloscope. Your oscillogram

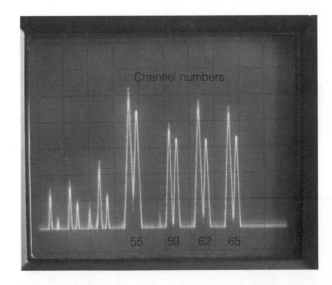

Figure 10.11 The four channels, shown on the screen of a spectrum analyser.

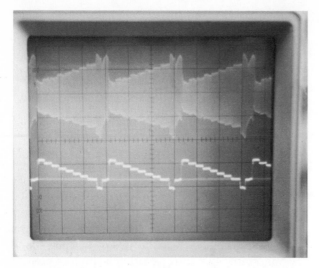

Figure 10.12 Top trace: the IF input to the vision demodulator. Bottom trace: the video signal output from the vision demodulator.

should now look something like the photograph in Figure 10.12.

3 Obtain information from Table 10.1 and Appendix 1 and draw a transmission spectrum. Show all relevant frequencies, such as the vision carrier, the sound carrier, and the vestigial sideband cut-off point. Show the spacing between the vision carrier and the sound carrier. State the following:

  i) the system type, e.g. B, G, I or D;
  ii) the type of vision modulation (positive or negative);
  iii) whether the sound carrier is FM or AM.

A simple television receiver

Figure 10.13 shows the block diagram of a simple television receiver.

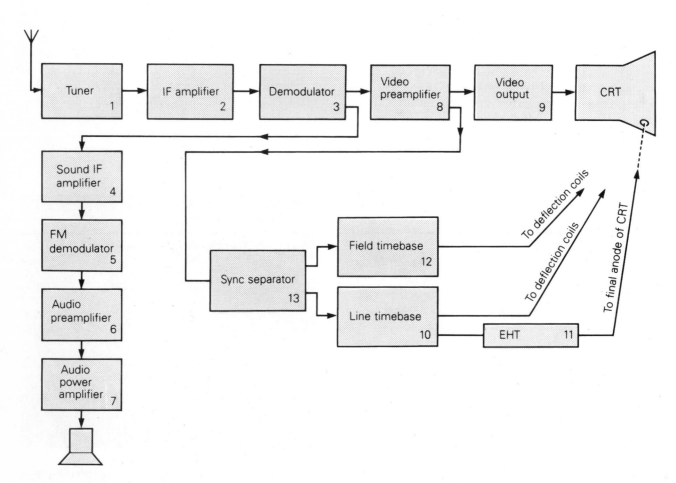

Figure 10.13 Block diagram of a simple television receiver.

Tuner (block 1)

The tuner consists of the RF amplifier, mixer and oscillator stages. Its job is to select the required channel by means of mechanical or electronic pushbuttons. The RF amplifier within the tuner amplifies the wanted signals. These signals are then mixed with a locally generated oscillator signal to produce two IF signals.

This method of operation, using the **superheterodyne principle**, simplifies amplification and gives excellent stability and good selectivity (see Chapter 5). As already mentioned, the sound and vision signals from the transmitter are based on different carrier frequencies. Therefore the output from the tuner produces two different IF signals. In the UK 625-line system I, the vision IF is 39.5 MHz and the sound IF is 33.5 MHz. (Note that the spacing between them is still 6 MHz.)

IF amplifier (block 2)

At least four stages of IF amplification are used. This represents about 80 dB of gain. The IF stages are usually controlled by some form of AGC, which prevents overloading from strong signals and also increases the IF gain for weak signals (see Chapter 6).

Demodulator (block 3)

The demodulator produces two outputs from these IF signals. One is the original modulating video waveform, the other is a second IF, which is at a frequency of 6 MHz. The latter is produced by beating together the sound and vision IFs (39.5 and 33.5 MHz) to produce a difference frequency of 6 MHz: the **intercarrier sound** signal. This 6 MHz 'beat' signal carries the original sound modulation, and it can be picked off just before or just after the video preamplifier. It is then fed to its own separate 6 MHz IF amplifier.

Sound IF amplifier (block 4)

Usually three or four stages of amplification are used here. The bandwidth of these amplifiers is about 220 kHz, centred on a frequency of 6 MHz. As the intercarrier signal is frequency modulated, these amplifiers are not usually controlled by AGC. Intercarrier sound is a receiver function and will be explained in greater detail in Chapter 18.

FM sound demodulator (block 5)

A **coincidence detector** is normally used to perform sound demodulation, but in older receivers a **ratio detector** may have been used. What follows on from this point is audio signal amplification (preamplifier, block 6), and then a power amplifier (block 7), which feeds the loudspeaker.

Video preamplifier (block 8)

Demodulated video is fed to a preamplifier or **buffer** stage. The preamplifier is usually included to prevent the video output stage (block 9) from loading the vision demodulator (block 3). The video preamplifier is designed so that the input impedance is high and the output impedance is low. This maintains a good match between the vision demodulator and the video output stage.

Video output stage (block 9)

The job of the video output stage is to drive the CRT display. It is primarily a voltage amplifier, which has to provide enough gain to drive the CRT cathode fully. This will then give a wide range of picture contrast. For a 20 in (50 cm) CRT, a gain of about 45 (about 33 dB) would be required. Also, this stage must have a wide bandwidth to give the required definition: usually 0 Hz – 5.5 MHz is aimed at.

Line timebase (block 10) and EHT (block 11)

The line timebase scans the CRT beam in the horizontal direction. It operates at a frequency of 15 625 times a second (15 625 Hz). This gives a single line period of $64\,\mu s$ (1/15 625 Hz) for scan and flyback. The line timebase can also be used to produce additional supply line voltages for different stages of the receiver. Bear this in mind if a fault occurs in this area: if the line timebase stops functioning altogether then these additional supply voltages will disappear.

The EHT generator is shown as a separate block, but in a real circuit it is part of the line timebase. The line timebase, because of its operating speed, is an ideal stage for developing the EHT voltage that is needed for the final anode of the CRT.

Field timebase (block 12)

The field timebase scans the beam in the vertical direction. It operates at a speed of 50 times a second (50 Hz). Field scan and flyback take 20 ms to complete.

Synchronising (sync) separator (block 13)

As already mentioned, the studio and the receiver images must have the same relative positions: the beam in the camera and the beam in the receiver CRT must be in the same position. Synchronising pulses are included with the video signal to enable the two sets of timebases to run in synchronism. The job of the sync separator is to separate the sync pulses from the video signal and then to divert them to their respective timebases: the line sync pulses to the line timebase and the field sync pulses to the field timebase.

ACTIVITY

4 For this activity you will need the circuit diagram of a monochrome television receiver. From information given within the circuit, construct a block diagram. Start with the tuner and continue with the IFs. Label all blocks to show their function: for example, video driver, video output, field timebase. Don't forget to show signal paths. Include additional information, such as transistor and integrated circuit reference numbers. You can write these numbers inside the blocks. Include any other details that you consider to be important. For help with this activity use Figure 10.13.

CHECK YOUR UNDERSTANDING

● When a carrier is modulated with a single modulating frequency, the carrier and two sidebands are generated from the modulator.

● System I video frequencies extend from d.c. to 5.5 MHz, and system G and B from d.c. to 5 MHz.
● Double-sideband transmission is not used for television transmissions because it is wasteful of bandwidth. Single-sideband transmissions cannot be used because of the difficulty of filtering close to the carrier.
● Vestigial sideband transmission is now universally used. It is a compromise between single sideband and double sideband. For television system I, channel bandwidth is 8 MHz. For system B it is 7 MHz.
● For system I, the sound carrier is higher than the vision carrier by 6 MHz. For system B, it is higher by 5.5 MHz. All systems except L and E use FM for the sound carrier.
● The spacing between the two carriers will depend on the system used (Table 10.1).
● Two types of amplitude modulation can be used for the video signal: negative and positive vision modulation. Negative vision modulation is now used almost everywhere throughout the world because of its superior advantages in coping with amplitude interference.
● There are three basic formats used for the transmission of colour television signals: PAL, SECAM and NTSC.

REVISION EXERCISES AND QUESTIONS

1 Why is negative vision modulation preferred to positive vision modulation?
2 With reference to the transmission of the vision carrier, why is it not advisable to use single-sideband transmission?
3 Why is it necessary to include a buffer stage between the vision demodulator stage and the video output stage?
4 How many scanning lines are there in 1 second for the UK 625-line system I?

Principles of colour television

Introduction

This is the first of three chapters that deal with the principles of colour television. We start this chapter with a brief look at the history of colour television, from its conception in 1921 to a partial colour television service in the UK in December 1967. We remind ourselves of the different colour television systems in use throughout the world, and compare the advantages and the disadvantages of each type. We then introduce the chromaticity diagram and the colour triangle, and mention the ideas of luminance and chrominance signals.

Any colour television system must incorporate receiver compatibility: that is, will both a colour receiver and a monochrome receiver function correctly using the same transmitted signal? This important topic is explained in this chapter, and again when we introduce a simple camera system.

In Chapter 12 we look at the way in which the two colour difference signals (the colouring signals) are derived from the red and blue camera outputs. We then explain why the chrominance signals are reduced in amplitude before they are added to the luminance signal and both signals modulate the vision carrier. Finally, we introduce balanced modulators and construct the composite chroma signal using Pythagoras' theorem.

The burst signal is explained in Chapter 13. We then introduce the idea of using a phasor to represent the composite chroma signal for a typical colour bar test pattern using six colours (colour clock).

History

In the 1940s and 1950s the worldwide growth in television receivers caused a recession in cinema audiences. The film companies and their producers responded to this by making more and more films in colour. Colour films and colour television pictures bring vitality; they are more real; the use of colour brings them to life.

In 1928, John Logie Baird (see Chapter 9) demonstrated a rather crude system of colour television using coloured discs in his system, which of course used mechanical scanning. Meanwhile, in the USA the Bell Telephone Company was demonstrating its own system, which also used mechanical scanning.

Also in the USA, a man called Goldmark produced his system in 1940. Goldmark's system was developed further by the Columbia Broadcasting System (CBS), and in 1951 an attempt was made to commission a colour television service. But like Baird's original monochrome television equipment, the CBS system still used mechanical scanning.

Discussions took place, and it was finally agreed to drop the mechanical system of scanning in favour of a new all-electronic system devised by the National Television Systems Committee (NTSC). The NTSC gave recommendations for a new 525-line compatible colour television service for October 1953. Regular transmissions commenced on 1 January 1954. Many other countries, including Japan, subsequently adopted the same system.

UK and Europe

In England in the early 1960s the British Broadcasting Corporation (BBC) carried out extensive tests using experimental transmissions. The system tested was similar in format to the NTSC's, but used 405 lines instead of 525 lines. Later it was decided to use 625 lines and to transmit the signal in the UHF band of frequencies. Comparative tests were also made in the years 1963–1964 using the French system (SECAM) and the German system (PAL). These are derivatives of the NTSC system, the PAL system being the closest.

SECAM stands for 'séquentiel couleur à memoire'. This system was proposed by Henri de France in 1958. PAL stands for 'phase alternation line'; the system was developed by the German Telefunken company under the leadership of Dr W. Bruch, who proposed the system in 1962. SECAM, PAL and NTSC are all capable of giving high-quality pictures; each system has its own advantages and disadvantages.

Comparing NTSC, PAL and SECAM

Under good signal conditions all three systems are capable of giving excellent results when viewed at a distance of 6–7 times picture height.

The NTSC system gives good vertical and horizontal resolution. The noise performance is also quite good, and the signal is relatively easy to transmit. The main disadvantage is that the colour signal can be prone to a type of distortion known as **differential phase distortion**. The chroma (colour) signal can be distorted in phase: that is, an unwanted phase shift occurs somewhere between the transmitter and the receiver, most likely in the receiver.

In the NTSC receiver, the chroma signal subcarrier uses a system of phase and amplitude modulation. The phase determines the colour transmitted, and the amplitude determines the saturation depth or purity of the colour on the screen. Therefore an unwanted phase change in the chroma signal results in a change of colour on the screen. A change in amplitude of the chroma signal corresponds to a change in saturation. For example, a reduction in amplitude could cause a colour that is supposed to be red to change to pink.

The main difference between PAL and NTSC is that the PAL system provides good immunity against differential phase distortion. The signal is a bit more difficult to transmit, in that the colour signal is itself more complex. The receivers are also more complex, and tend to be a bit more expensive. However, recording the signal onto tape presents less of a problem. The PAL signal also suffers less when transmitted over long distances, which is not the case for NTSC. The main disadvantage of PAL is slight incompatibility when viewing a colour signal on a monochrome receiver. The chrominance horizontal resolution or definition is also reduced slightly compared with NTSC.

The SECAM system also suffers less from differential phase distortion. The receivers tend to be less complex when it comes to decoding the colour signal. Again, recording the colour signal onto tape presents minimum problems. The biggest disadvantage with the SECAM

system is that the chrominance resolution (fineness of detail) is reduced, because one half of the chrominance signal is discarded prior to transmission.

With good-quality transmissions and well set-up receivers, it is difficult to distinguish which system is being used. In 1966 the UK TV Advisory Committee proposed that the PAL system should be adopted for the UK. Limited hours of broadcasting commenced in July 1967, and in December 1967 this was made official.

Chromaticity

> The study of colour is called **colorimetry**. Information concerning the hue and saturation of a colour is called its **chromaticity**.

The **CIE chromaticity diagram** shown in Figure 11.1 gives precise information about dominant wavelength and purity. Originally produced by an International Commission on Light in 1931, it is still used today as the standard reference. It is a kind of colour map on which colours can be pinpointed by using a pair of coordinates.

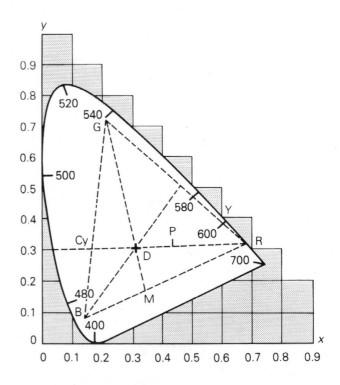

Figure 11.1 The CIE chromaticity diagram.

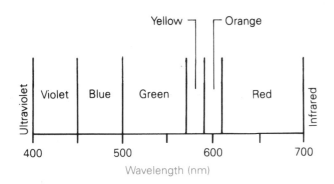

Figure 11.2 Approximate distribution of colours in the spectrum.

Figure 11.2 shows how the colours are distributed in the spectrum.

The chromaticity diagram is shaped like a tongue. The wavelengths of the various colours are marked off around the edge. The colours therefore follow the sequence given in Figure 11.2, from 400 to 700 nm (nm = nanometres).

Starting at the bottom left-hand corner and working clockwise, the colours on the edge of the diagram represent the pure colours: that is, 100 per cent saturated. From any point on the edge to the white reference point marked D, the colour gradually becomes paler or less saturated.

The nature of white light can also vary. For example, the light from a tungsten lamp might be thought of as white at night-time, while the light obtained from a fluorescent tube might be 'daylight'. The CIE reference white that is used for colour television is called **illuminant D**. It can be found by using the coordinates $x = 0.313$ and $y = 0.329$.

We can specify the exact hue of a colour from the chromaticity diagram. A line joining point D to the edge of the tongue 'cuts' the dominant wavelength of the particular hue or colour. In this example the line CY cuts the edge at 492 nm.

We can also specify the purity of a colour. Look at the line that crosses D and continues through P to R. The red becomes paler as we move from the edge of the diagram at R to the centre D. At P, the amount of saturation is only one third of what it is at R. More accurately, the purity is 33 per cent.

If we project coloured lights onto a white screen, we can obtain various coloured mixtures by varying their proportions. This **additive mixing** process is that used on the screen of a colour television receiver tube. Red, green and blue lights provide the greatest combination of colours for practical purposes. This is the reason for choosing these three **primary colours** for use in colour television. Do not forget that we can produce white by mixing together the correct proportions of red, green and blue light (Figure 11.3).

Red and green light of equal intensity light give yellow. Therefore if yellow and blue are mixed correctly we get white. Blue is the **primary** colour and yellow is the **complementary** colour. The same is true for the pairs red and cyan, and green and magenta. You need to memorise all of the above primary and complementary colours and their resultants, because you may need to recognise which colour, if any, is missing from a received picture.

We can use the chromaticity diagram to see the result of mixing coloured light. For example, if we draw a straight line between the two points R and G, we can see along this line the colours that can be obtained by mixing red and green lights in different proportions. If RY equals GY – that is, red 50 per cent and green 50 per cent – we get the colour yellow. If we now take the point at 600 nm, then red is about 90 per cent and green is about 10 per cent. The colour obtained is orange.

The red end and the violet end of the spectrum cannot come together, but if these colours are mixed we get the purple shades. These have no specified frequency because they are not natural colours. The chromaticity diagram gives the mixture along the bottom straight line between 400 and 700 nm.

In 1931 the CIE adopted for use three dominant wavelengths: red (700 nm), blue (435.8 nm) and green (546.1 nm), all at 100 per cent purity. If we join these three points together on the chromaticity diagram we get a triangle, which embraces all the colours that can be reproduced from these three primary colours.

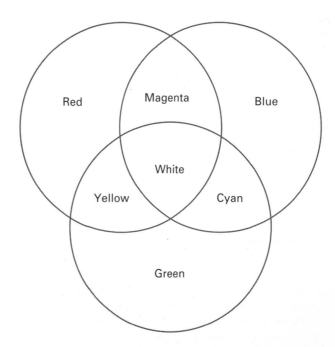

Figure 11.3 The principle of mixing coloured light.

The colour triangle

We have already mentioned the colour triangle with reference to the chromaticity diagram (Figure 11.1). The corners of the triangle contain the primary colours, and the result of mixing any two of them is shown along the appropriate side. The centre of the triangle is white where all the three lights or colours are present, if they are in the right proportions.

The primary colours used for colour television are determined by the fluorescence of the available phosphors used in the CRT, as it is the tube that gives the colour. Consequently, the primaries used by the NTSC for television purposes are not the same as the CIE primaries. The NTSC primaries are shown on the chromaticity diagram, plotted from the tabulated coordinates.

The television receiver can produce any colour inside the triangle. The coverage is better than that obtainable by colour photography or printing. The colours outside the triangle cannot be reproduced, but the eye does not miss them, as they are rarely apparent in everyday scenes.

The disadvantage of the colour triangle is that, unlike the chromaticity diagram, it cannot represent all colours.

ACTIVITIES

1   Obtain three electric bulbs: red, green and blue. They can be the type used in a flashlight, but three mains-operated bulbs would be ideal. If you are using mains-operated coloured bulbs, then you need to obtain three cylinders about 45 cm long, one for each bulb. Fit the cylinders over the bulbs to help to channel the light into a round spot.

     First, shine the red and green lights onto a white surface. Overlap the two light beams. Where they overlap, you should see the colour yellow. Next, do the same with the blue and red lights. In this case you should see the colour magenta. Finally, try the blue and green lights. You should now see the colour cyan (a bluey green). If you overlap all the light beams, you should produce a type of white, depending on the intensity and colour of each individual bulb.

2   For this activity you will need a colour television receiver. Any type will do – NTSC, PAL or SECAM – that is, the type of television used in your country. You will also need a pattern generator capable of supplying a full colour bar pattern (most of them do), and a magnifying glass.

     Set the pattern switches on the generator to 'colour bars', and tune in the television receiver to the resulting pattern. Adjust the contrast, brightness and colour controls for a normal picture (ask your tutor to help you with this).

     Using the magnifying glass, look closely at the white bar. You should see that the white bar is made up of red, green and blue dots or – if the tube is a more modern type – small rounded rectangles. Now, without the magnifying glass, look at the white bar. What you should see now is that the eye integrates or mixes the three separate colours, and gives the impression of white.

     Repeat this with the yellow, magenta and cyan bars. You should see that the yellow bar is made up from red and green dots; the magenta bar from red and blue dots; and the cyan bar from blue and green dots. For the three primary coloured bars, only those dots are illuminated by their respective electron beams: red, green or blue. For black, none of the dots are illuminated.

3   For this activity, you need the same equipment as in Activity 2. Turn the colour control to minimum. The colour bars should turn into different shades of grey. Now inspect each bar in turn using the magnifying glass. What do you see? Each grey bar (not the black) is made up of equal amounts of red, green and blue dots. The shade of grey that represents the yellow bar is not as bright as the white one. The shade of grey that represents cyan is not as bright as the yellow bar. This continues until the black bar is reached. This shows that a colour television receiver is fully capable of showing any shade of grey, black or white picture.

The apparatus for activity 1 should be constructed by a qualified person, and made safe to use. If you are using mains-operated lamps, it is important to use the correct wire and lampholders, and to wire them up correctly. Take care when you insert the lamps into the cylinders. Don't use cylinders made of a material that could get hot or, even worse, melt.

Luminance and chrominance signals

Now, for our purpose, the word **acuity** means 'sharpness of vision'. Acuity is a measure of our eyes' ability to observe fine detail. Tiny objects or fine detail are not visible if we stand well away from them. The great impressionist painters exploited this fact very well. The paintings show no detail, but they appear at a distance to be complete. However, closer inspection shows them to be very rough indeed. If we view a television picture at a reasonable distance, the line structure of the raster disappears: that is, we cannot see the individual lines that go to make up the raster.

If a small coloured object, such as a thread of cotton, is moved away from an observer, its colour will eventually be lost, although it will still be seen as a grey object. The shade of grey will depend on the colour of the object. Observe an aircraft in the sky: you may see the aircraft but you probably won't be able to tell the colour of its insignia. Some colours are lost more quickly than others. For example, blue soon appears as a shade of grey. Red is more noticeable.

What all this means to us as television viewers is that there is no point in transmitting fine detail in colour television signals, if that detail would be seen only as different shades of grey. In fact, in the PAL system, the maximum definition that is used for the chroma or colour signal corresponds to a frequency of about 1.0 MHz. As already mentioned earlier, the maximum bandwidth allowed for the video signal is 0–5.5 MHz. What is actually broadcast from the television transmitters is a very high-definition video or **luminance** signal with a low-definition colour or **chrominance** (chroma) signal. The complete colour picture then appears in full detail. The human eye's lack of acuity is used to advantage, and channel space is saved with respect to the colour signals.

A colour picture on a colour television tube is made up of light emitted by a large number of tiny red, green and blue dots. The amount of light from each dot will determine the colours viewed from a normal viewing distance. As we have already seen, our eyes are not capable of seeing the small separate dots of colour, but only the additive mixing of all three primary colours.

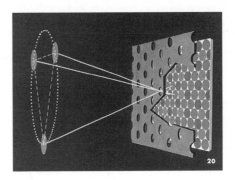

Figure 11.4 The inside of a colour CRT, looking towards the screen area.

The shadowmask cathode ray tube

Figure 11.4 shows the inside of a colour CRT, looking towards the screen area. Inside the glass envelope there are effectively three cathode ray tubes. There are three electron guns, each associated with one of the three screen phosphors. The phosphors fluoresce in the primary colours: red, blue and green. They are deposited onto the face plate with great accuracy, in rows of dots or rectangles. Each phosphor screen contains about 900 000 dots, which are so small that they are not separately visible at normal viewing distance. The eye integrates them into a large single colour screen. The three sets of dots are mixed to form three screens in one.

Figure 11.5 shows that the beams are fired from their own electron guns. The guns are tilted slightly towards the CRT axis. The three beams converge as they enter the hole in the mask. We shall discuss the colour television picture tube in more detail in Chapter 21.

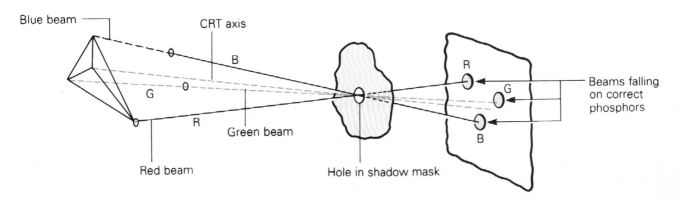

Figure 11.5 Arrangement of the three beams in a CRT.

ACTIVITIES

4   You can confirm the acuity of the eyes by holding up different pieces of coloured cotton. Get someone to help you with this. Walk away from the cotton threads and see if you can still recognise the colours. Beyond a certain distance some of the colours will turn to a shade of grey. This shows that the human eye is not very sensitive to colour detail.

5   For this activity you will need the assistance of your tutor or someone in a responsible position. You will also need a colour television receiver. Tune the television to a noise-free colour channel. Inspect the service manual, and look for a way of preventing the luminance signal from reaching the final matrix circuit. This will result in only the chrominance signal being applied to the tube cathodes. You should now notice that you can see patches of colour on the screen, but no luminance information. The chrominance signal should be of low definition. This shows that the luminance signal is also needed for a colour television receiver.

Compatibility

The most basic requirement of any colour television system adopted for public service is that it should give a satisfactory picture on a monochrome receiver: that is, the colour signal must be **compatible**. Also, the colour receiver must give a satisfactory black-and-white or monochrome picture when it is tuned to a monochrome transmission. This is known as **reverse compatibility**. These features are essential if existing systems of transmission and reception are to be fully used.

Compatibility and reverse compatibility are achieved because colour and monochrome transmissions use the same basic standards (see Chapter 10). The colour transmission contains extra information, which can be interpreted by the colour receiver only. In a monochrome receiver, the colour transmission is non-effective: the receiver takes no notice of it. It uses the luminance signal only, and from it produces a black-and-white picture.

> All colour television systems must be compatible with monochrome receivers.
> All colour television receivers must work from a monochrome transmission.

A monochrome TV transmission does not activate the colour circuitry in a colour receiver, but the basic luminance signal is used and passed on to the cathodes of the CRT to provide a black-and-white picture. For a colour receiver to show a complete colour picture, it needs the luminance signal *and* the chrominance signal.

ACTIVITY

6   For this activity, you will need a colour television receiver, a colour bar generator, a dual-trace oscilloscope, and high-impedance oscilloscope leads.

Tune the colour receiver to the colour bar signal and adjust the receiver controls for a normal colour bar picture. Connect the oscilloscope lead to the red, green and blue cathodes in turn. Set the oscilloscope timebase to $10\,\mu s$/division to show one line of drive signal for each of the three cathodes. The waveform on the oscilloscope screen should be upside down: that is, inverted. Figure 11.6 shows the result for the red cathode. From these different waveforms, determine the active part of the picture line, in microseconds.

All three waveforms in turn are made up from the Y' or luminance signal and each colour difference signal. For example, the red drive signal is made up from the Y' signal and the $R - Y'$ signal:

$$R = (R - Y') + Y'$$

Similarly for the green:

$$G = (G - Y') + Y'$$

and the blue:

$$B = (B - Y') + Y'$$

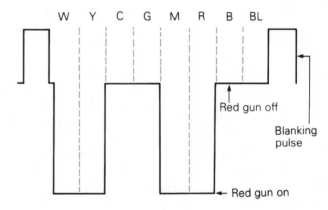

Figure 11.6 One cycle of the drive signal for the red cathode.

(These terms are explained later.)

While viewing one of the cathode drive signals on the oscilloscope, turn down the receiver colour control so that a grey scale appears on the screen. What happens to the waveform on the oscilloscope? The colour difference signals disappear and leave behind the luminance signal only. This effect should be similar for each of the cathodes.

The colour camera

As we have seen, a monochrome camera contains:

1. a system of lenses for bringing the studio scene into focus;
2. the camera tube photosensitive plate, which generates the video signal voltage;
3. some sort of preamplifier, which amplifies the minute signal voltage to a suitable level.

The camera output voltage is usually set for about 1 V p.p. when looking at peak white.

A colour camera contains all of the above, three times over. It also has to have some form of colour-splitting arrangement fitted between the lens system and each of the three camera tubes. The problem with using multiple tubes is the need for each picture component to register accurately with the other two. A point in any one picture must be scanned at exactly the same instant by the other beams in their own tubes. This is especially important for the luminance signal, as misregistration will cause blurred edges on the image.

Figure 11.7 shows the idea of using three camera tubes

to develop the necessary signals. Also shown are the **dichroic mirrors** ('dichroic' means 'two colours') and special light filters. These adjust the response of the camera for each of the primary colours. Although this scheme is now obsolete, it provides a useful study of colour splitting.

In Figure 11.7 the **camera lens**, which may be one of a set usually mounted on a rotating turret, reproduces the scene. Behind this is mounted a **field lens**. The job of this lens is to converge the image into the **relay lens**. The relay lens simply causes the focused image to be relayed from the field lens to the photosensitive layer of each tube.

Because of the action of the dichroic mirrors, the camera tubes respond only to red, blue or green light. The action of the mirror is very complex, but its effect is to transmit (let through) one end of the colour spectrum – red, for example – and reflect the other end (in this example, blue). There is a 'crossover' region between the transmission and reflection wavelengths, and this is used to give green reflection or transmission.

Figure 11.8 shows a commonly used colour splitter arrangement. A lens assembly provides suitable focusing of the studio scene. DM1 is a blue-reflecting dichroic mirror. The mirror transmits all the red end of the spectrum and about 90 per cent of the green. The other 10 per cent of the green is reflected with the blue onto a surface-silvered mirror RM1, and then to the blue camera tube.

DM2 is a red-reflecting dichroic mirror, which reflects all the reds transmitted by DM1 onto a second surface-silvered mirror, RM2, and then to the red camera tube. Some of the green is also reflected via this dichroic mirror (DM2), but more than 80 per cent of the green content is passed on to the green camera tube.

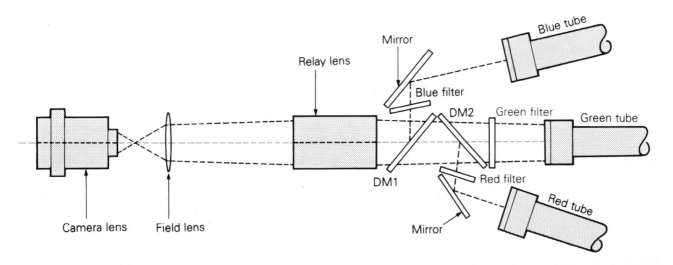

Figure 11.7 A three-tube camera system, showing dichroic mirrors and light filters.

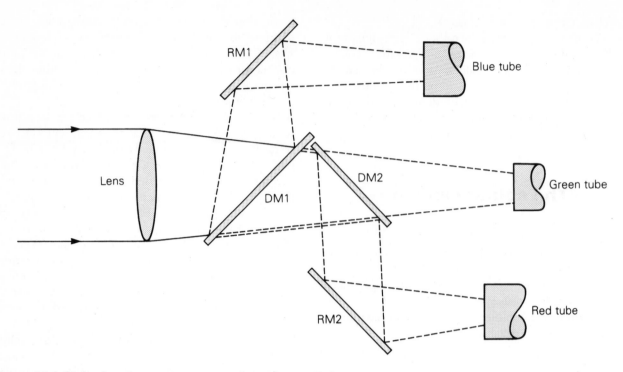

Figure 11.8 Dichroic mirror arrangement for colour splitting.

Figure 11.9 shows the response curves of the three camera tubes. Each one has a peak response at the correct primary colour wavelength. The response curves must also 'overlap' their neighbouring colours. The precise shaping of the response curves depends upon the quality of the dichroic mirrors, the colour correction filters and the camera tube spectral response.

The dichroic mirror system has now been replaced by a single dichroic prism.

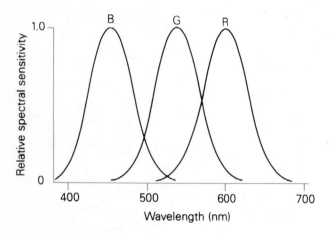

Figure 11.9 Spectral responses of the colour camera.

CCD imagers

Most modern studio and portable cameras (camcorders) in use today are of the CCD (**charged-coupled device**) type. They utilise solid-state, light-sensitive units, which are far superior to the conventional vidicon or image orthicon type of tube, for the following reasons:

1. They are very robust. This is important when they are used in portable camera equipment such as camcorders.
2. They suffer less if they are exposed to overbright lights.
3. They produce good pictures with a high signal-to-noise ratio.
4. **Image lag** (the 'ghosting' effect caused by poor lighting conditions) is less of a problem.
5. They do not suffer from ageing, or from developing the condition known as **low emission**.
6. They suffer less from performance drift (that is, any drift in performance due to such things as temperature change or component tolerances).

The latest information from the BBC indicates that, based on the number of 'tubed' cameras still in use, roughly 95 per cent of colour cameras are now CCD based.

Gamma correction

Gamma correction was mentioned in Chapter 10, in the context of non-linearity within the camera tube. Unfortunately, in electronics, very little is linear. This non-linearity also exists within the CRT in the receiver. To allow for it, corrections are made to the signals that are generated from the camera system.

Suppose a cathode ray tube in a receiver needs 60 V between the grid and the cathode for cut-off (that is, black level), and 0 V for maximum brightness or white level. Then for 30 V bias one might expect a mid-grey, 45 V bias would give dark grey, and 15 V bias would give light grey: in other words, a uniform change of bright ness with uniform changes in grid/cathode voltage. Unfortunately, this is not what happens in practice. The relationship between the light the output from the CRT and the bias is not linear.

Figure 11.10 shows a characteristic curve of a CRT. Light output L is plotted against negative grid voltage V_g or bias. The curve is similar to the I_a/V_g curve of a thermionic valve. The curve is approximately parabolic in shape, because it almost follows a square law. The actual power of the independent variable is not actually 2, but lies between 2.2 and 2.7. This figure is referred to as the **gamma** of the tube. That is:

$$L \propto V_g{}^\gamma$$

(γ is the Greek letter gamma.)

If we look at the characteristic curve we can see that for a bias of 30 V we do not get mid-grey but some darker shade of grey. Except for black and white, this is the case for any bias. The light output from the tube is less than it would be for a linear characteristic. The ideal transmitted signal output is indicated by the points A, B, C, D and E for the different brightness levels. If these voltage changes were fed to the CRT in the receiver, then the light output from the tube would follow the points A′, B′, C′, D′ and E′. The only illumination levels that are correct are white and black. The other illumination levels are all low, and hence incorrect. They should ideally follow the points b′, c′ and d′. These are projected from the **ideal characteristic curve**, which is of course a straight line.

To overcome this brightness distortion, the transmitted signal output must be **gamma corrected**. This means that the brightness levels follow the curve A, b, c, d, E. These points intercept the actual tube characteristic in such a way that the output follows the ideal response A′, b′, c′, d′, E′.

Gamma correction is performed at the transmitter before the colour difference signals are formed, and before the matrix circuit (see Chapter 12). For monochrome transmissions, gamma correction is still necessary, otherwise an incorrect grey scale would result.

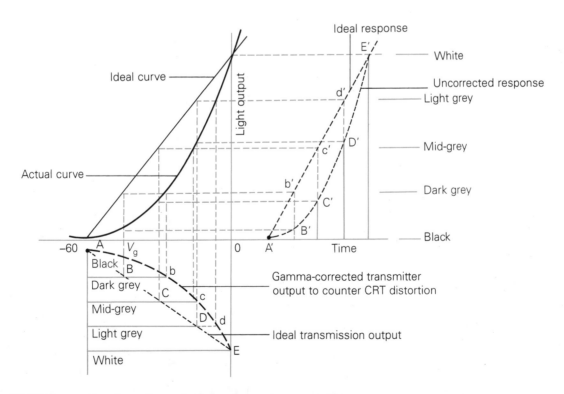

Figure 11.10 Curves to show the principle of gamma correction.

A signal voltage that has been gamma corrected is usually indicated with a 'prime'. For example, E'_G would indicate a corrected 'green' camera output.

> Gamma correction is used at the transmitter to overcome the inherent non-linearity of both the camera system and the receiver CRT.

■ CHECK YOUR UNDERSTANDING

● There are three types of terrestrial television system: PAL, NTSC and SECAM. PAL and SECAM provide immunity against differential phase distortion.

● Colorimetry is the study of colour. Information about its hue and saturation is known as chromaticity.

● White light can be produced by mixing equal intensities of red, green and blue light.

● The three primary colours used for colour television are red, green and blue. The complementary colours are yellow, cyan and magenta.

● The white chosen for colour television is known as Illuminant D.

● Acuity measures the sharpness of our vision: how well our eyes can see fine detail.

● Coloured objects at certain distances will appear as shades of grey.

● The luminance signal is of high definition. The chrominance signal is of low definition.

● Compatibility means that a monochrome receiver should perform well from a colour signal. Reverse compatibility means that a colour receiver should produce a good monochrome picture from a monochrome signal.

● The chrominance signal is transmitted within the existing bandwidth of the television signal.

● A monochrome receiver uses only the luminance signal from a colour transmission. The colour receiver uses both parts of the signal: the luminance and chrominance parts.

The PAL transmitter

The PAL encoder

Figure 12.1 shows the block diagram of a typical PAL encoder. Note that the three-tube camera that we discussed previously, complete with its lens, dichroic mirrors and filters, is now shown as a small block, labelled 'three-tube camera'.

The first job in our block diagram is to provide a way of obtaining the luminance signal. This is done by connecting the camera outputs to a circuit called a **matrix**. A matrix is usually applied in television when a number of signals are combined in their correct proportions, in order to get a resultant output signal. In our case we need to derive the luminance signal from the camera output signals. In this example we have used a simple resistor arrangement based on potentiometers (in a real encoder the matrix would be a lot more complex).

Before proceeding with this description of the PAL encoder, it is worth mentioning a suitable test picture that could be used for the camera. It is called the **colour bar display**.

The BBC transmits a test signal that is used to check the transmission and reception of colour television signals. The saturation level is 95 per cent with an amplitude of 100 per cent. This simply means that each colour is 95 per cent pure, with a dilution of white at 5 per cent. However, to simplify the explanation, let us assume that in the colour bar signal all primary and complementary colours will be at 100 per cent saturation.

The colour bar pattern is shown on the cover of this book. It has eight vertical bars of equal width: three primary colours (red, green and blue); three complementary colours (yellow, cyan and magenta); and white and black.

The whole pattern is arranged in descending order of luminance: that is, from left to right the luminance or brightness value for each colour reduces.

As we have already mentioned, the luminance signal must be obtained from the camera output signals: that is, from the matrix. The correct shade of brightness must correspond to each colour. For a luminance signal that will give the best overall white and any shade of grey, experience has shown that the best results are obtained by taking 59 per cent of the green camera output, 30 per cent of the red camera output and 11 per cent of the blue camera output. The proportions are unequal because the human eye does not see all colours with the same brightness level, or luminosity. It is more sensitive to green than to red, and even less sensitive to blue.

This can be shown using the **relative luminosity curve**, which is reproduced in Figure 12.2. It is basically a frequency response curve of the average human eye, and you will notice that the eye is most sensitive to paler green. This has a wavelength of about 555 nm. This colour is given a value of 100 per cent. If the human eye 'sees' the relative brightnesses of the different colours like this, then any television system must present them in the same way.

Early monochrome television cameras suffered from being too sensitive at the red end of the spectrum. The result was that the shade of grey produced by the camera looking at a red object was not correct. Special make-up was used for the actors or presenters 'on camera' to overcome the problem and produce the correct shade of grey.

Assume that the camera in Figure 12.1 is pointing at the colour bar picture, which contains the six colour bars plus the white and black bar. Therefore all 575 picture lines produced from the camera system will look the same. The camera outputs at R,G and B are adjusted for 1 V peak for a white scene.

Let us consider the white bar. Points R, G and B will be at 1 V peak. What about points a, b and c? Clearly, because of the position of the sliders on the potentiometers, the voltages at these points will not be at 1 V. In fact point a will be at 0.59 V, point b will be at 0.3 V and point c will be at 0.11 V. These three separate voltages are then added together in the 'add' circuit.

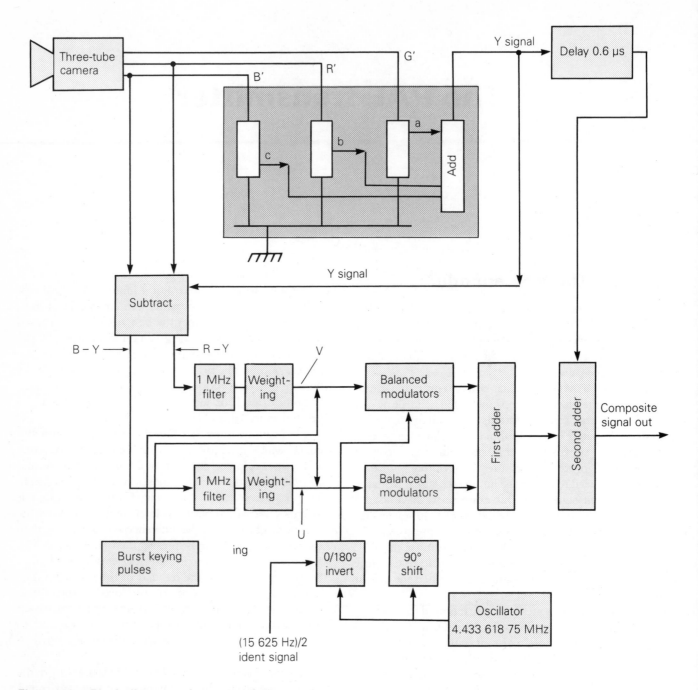

Figure 12.1 Block diagram of a typical PAL encoder.

Expressed mathematically the luminance signal is equal to 0.59G′ + 0.30R′ + 0.11B′. The symbol for the gamma-corrected luminance signal is Y′. Therefore, for the white bar, the luminance value is given by

$$Y' = (0.59 \times 1\,\text{V}) + (0.30 \times 1\,\text{V}) + (0.11 \times 1\,\text{V}) = 1\,\text{V}$$

Note that if normal symbols were being used, the luminance equation would be written as

$$E'_Y = 0.59\,E'_G + 0.30\,E'_R + 0.11\,E'_B$$

where E′ means that the voltage has been gamma corrected, and the subscripts Y, G, B and R indicate which voltage is being considered.

For simplicity, the equation is commonly used in the form

$$Y = 0.59\,G + 0.30\,R + 0.11\,B$$

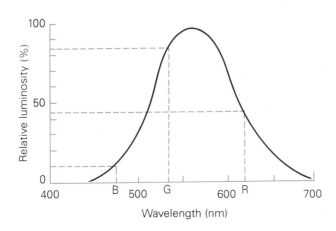

Figure 12.2 Relative luminosity curve.

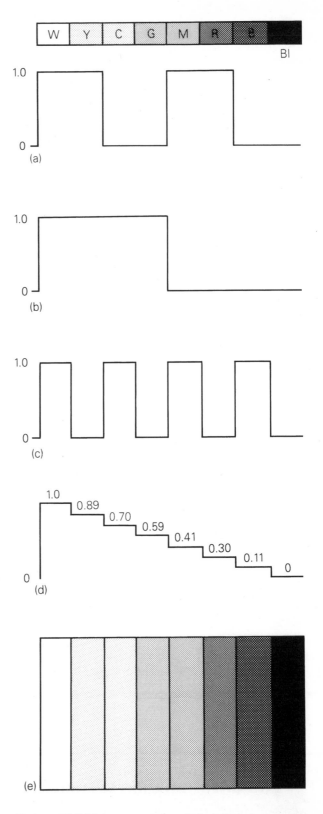

The prime is usually omitted unless a specific point is being made. However, remember that Y, G, R and B represent varying voltages in this context. They are not unit symbols, unlike the V in 230 V, which stands for volts (the unit of measurement). From now on in this discussion we shall drop the prime and assume that all values are gamma corrected.

For the yellow bar, there is no blue, so the blue camera output will be zero. The red and green camera outputs (R and G) will be at 1 V each. So the luminance value for this colour bar will be equal to 0.59 of green added to 0.3 of red:

$$(0.59 \times 1) + (0.30 \times 1) = 0.89 \, V$$

which is the luminance value for yellow.

For the colour bar scene, the red, green or blue camera outputs will be at either 0 V or 1 V depending on the colour in view. For example, for yellow the red and green camera outputs will both be at 1 V, or 100 per cent. The blue will be at 0 V, or 0 per cent.

Figure 12.3 shows the voltage output of the camera tubes at points R, G and B in Figure 12.1. The **red** camera output (Figure 12.3a) produces 1 V output at point R for the white, yellow, magenta and red bars, and 0 V output for the cyan, green, blue and black bars. The **green** camera output (Figure 12.3b) produces 1 V output at point G for the white, yellow, cyan and green bars, and 0 V output for the magenta, red, blue, and black bars. The **blue** camera output (Figure 12.3c) produces 1 V output at point B for the white, cyan, magenta and blue bars, and 0 V output for the yellow, green, red, and black bars.

The **luminance** values for all of the colour bars are shown in Figure 12.3d. These values are derived from the formula given earlier for the luminance signal, as follows.

Figure 12.3 Voltage output of the camera tubes at points R, G and B in Figure 12.1: (a) red tube output; (b) green tube output; (c) blue tube output; (d) luminance steps.

For white:

$$Y = (0.30 \times 1.0) + (0.59 \times 1.0) + (0.11 \times 1.0)$$
$$Y = 1 \text{ V}$$

For yellow:

$$Y = 0.3(1) + 0.59(1) + 0.11(0) \text{ (no blue)}$$
$$Y = 0.89 \text{ V}$$

For cyan:

$$Y = 0.3(0) + 0.59(1) + 0.11(1) \text{ (no red)}$$
$$Y = 0.70 \text{ V}$$

For green:

$$Y = 0.3(0) + 0.59(1) + 0.11(0) \text{ (no blue or red)}$$
$$Y = 0.59 \text{ V}$$

For magenta:

$$Y = 0.3(1) + 0.59(0) + 0.11(1) \text{ (no green)}$$
$$Y = 0.41 \text{ V}$$

For red:

$$Y = 0.3(1) + 0.59(0) + 0.11(0) \text{ (no green or blue)}$$
$$Y = 0.30 \text{ V}$$

For blue:

$$Y = 0.3(0) + 0.59(0) + 0.11(1) \text{ (no red or green)}$$
$$Y = 0.11 \text{ V}$$

For black, the luminance value is zero.

As you can see, we have produced the Y or luminance signal from the camera system by taking the correct proportions of the outputs of each camera tube. Compatibility so far holds true. The monochrome receiver can use the Y signal to give a picture that will show six bars of different shades of grey plus a white bar on the left of the screen and a black bar on the right of the screen (Figure 12.3e).

> Monochrome receivers use the Y signal to show a black, white and grey version of the coloured scene.

The chrominance signals

To produce a colour copy of the original studio scene, the colour receiver needs extra signals. These signals are known as **chrominance signals** (the chrominance signals are not used by the monochrome receiver).

The complete or composite chrominance signal is produced from **two** separate chrominance signals. These give information about the **red** and **blue** content in the studio scene. Green chrominance information is not transmitted at all; it is discarded. It is used only to form the Y signal in the encoder. If we were to transmit information about the green scene content as well, we would need a greater bandwidth for the television channel. It would also make the complete system too complicated.

If we know the values of the luminance signal and the two chrominance signals (red and blue), we can obtain the green chrominance signal in the colour receiver. (We shall return to this subject in Chapter 20.)

The two chrominance signals (red and blue) give information about saturation and hue. **Hue** means colour. **Saturation** refers to how pure the colour is. If the colour red is less saturated then it turns into a shade of pink.

The two chrominance signals are obtained by subtracting the Y signal from the camera outputs of red and blue to produce what is known as **colour difference signals**. So, red minus the luminance signal is the $R - Y$ colour difference signal, and blue minus the Y signal is the $B - Y$ colour difference signal. The result of this process is that we now have two colour difference signals, one called $R - Y$ and the other one called $B - Y$.

We can now calculate the two colour difference signals. For this we shall again use the colour bar scene. We know that red or blue will either be at maximum value (1 V) or at minimum value (0 V). For the $B - Y$ signal, we subtract Y from the values of the blue camera output. For example, for white, blue is needed, so the blue camera output is 1 V. The luminance is also at a maximum: that is, $Y = 1$ V. So the $B - Y$ signal is $1 - 1 = 0$ V. Table 12.1 shows the results of this calculation for all the colour bars. Figure 12.4a shows the shape of the resulting signal for one chroma line.

For the $R - Y$ signal, we now subtract Y from the values of the red camera output. Table 12.2 shows the results of this, and Figure 12.4b shows the shape of the $R - Y$ signal for one chroma line.

Note, that for both sets of calculations, for white or black the $R - Y$ and $B - Y$ signals are both at zero. This is also true for any shade of grey (see later).

If the $G - Y$ signal had been used, it would have looked like the diagram in Figure 12.4c.

You may have noticed that it is possible for the two colour difference signals to have positive or negative values, whereas the luminance signal is always positive. It is essential that these positive and negative values are preserved during transmission and reception, as they are responsible for operating the receiver CRT *together* with the luminance signal. If the identity of these polarities were lost, the result would be incorrect colours.

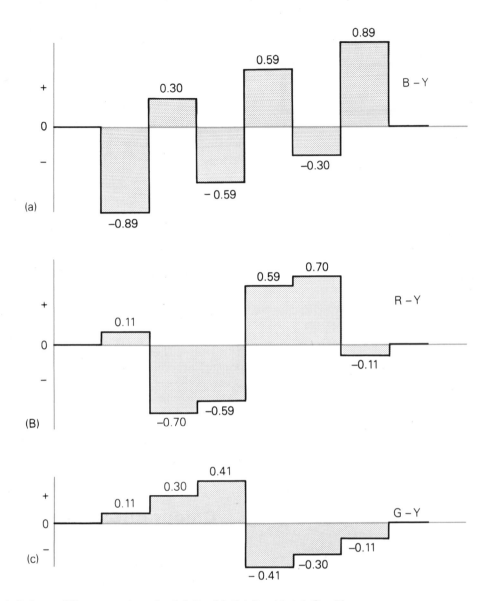

Figure 12.4 Colour difference signals: (a) B – Y; (b) R – Y; (c) G – Y.

Table 12.1 Derivation of the B – Y colour difference signal

	Blue camera output	minus	Luminance signal	=	B – Y
White	1		1.0		0.00
Yellow	0		0.89		–0.89
Cyan	1		0.70		0.30
Green	0		0.59		–0.59
Magenta	1		0.41		0.59
Red	0		0.30		–0.30
Blue	1		0.11		0.89
Black	0		0.00		0.00

Table 12.2 Derivation of the R – Y colour difference signal

	Red camera output	minus	Luminance signal	=	R – Y
White	1		1.00		0.00
Yellow	1		0.89		0.11
Cyan	0		0.70		–0.70
Green	0		0.59		–0.59
Magenta	1		0.41		0.59
Red	1		0.30		0.70
Blue	0		0.11		–0.11
Black	0		0.00		0.00

The colour difference signals can have positive or negative values. The luminance signal is always positive. For white or black in the scene, the colour difference signals are zero.

The luminance signal for grey scale

What happens to the luminance signal when the camera is looking at a black or white or grey scene? To answer this question, we shall assume that the camera is looking at a theoretical grey-scale equivalent of the colour bar scene:

Table 12.3 The green, red and blue outputs for a grey-scale camera scene

	Green output			Results	Row numbers
White	1	×	0.59	= 0.59	1
Very light grey	0.89	×	0.59	= 0.52	2
Light grey	0.70	×	0.59	= 0.41	3
Mid-grey	0.59	×	0.59	= 0.35	4
Grey	0.41	×	0.59	= 0.24	5
Dark grey	0.30	×	0.59	= 0.18	6
Very dark grey	0.11	×	0.59	= 0.06	7
Black	0.00	×	0.59	= 0.00	8

	Red output			Results	Row numbers
White	1	×	0.3	= 0.30	1
Very light grey	0.89	×	0.3	= 0.27	2
Light grey	0.70	×	0.3	= 0.21	3
Mid-grey	0.59	×	0.3	= 0.18	4
Grey	0.41	×	0.3	= 0.12	5
Dark grey	0.30	×	0.3	= 0.09	6
Very dark grey	0.11	×	0.3	= 0.03	7
Black	0.00	×	0.3	= 0.00	8

	Blue output			Results	Row numbers
White	1	×	0.11	= 0.11	1
Very light grey	0.89	×	0.11	= 0.10	2
Light grey	0.70	×	0.11	= 0.08	3
Mid-grey	0.59	×	0.11	= 0.06	4
Grey	0.41	×	0.11	= 0.04	5
Dark grey	0.30	×	0.11	= 0.03	6
Very dark grey	0.11	×	0.11	= 0.01	7
Black	0.00	×	0.11	= 0.00	8

that is, eight bars consisting of white on the left, black on the right, and different shades of grey in between (Figure 12.3e).

Recall the formula for calculating the luminance signal:

$$Y = (0.59\,G) + (0.30\,R) + (0.11B)$$

Table 12.3 shows another way of arriving at the values for the luminance signal as discovered earlier. To prove this, add together all rows with the same number. For example:

Green row number 2 is equal to 0.52
Red row number 2 is equal to 0.27
Blue row number 2 is equal to 0.10
Added together this gives: 0.89

You will have noticed that the camera outputs at points R, G and B decrease with correspondingly darker shades of grey. Maximum output for the white bar and zero output for the black bar. Put simply, a light grey cannot produce as much signal at the outputs of *each* of the camera points. Therefore points R, G, and B reduce with each darker shade of grey. For black they will all be at zero. Again, compatibility still holds good, and Figure 12.3d still appears at the output of the matrix.

The monochrome receiver uses the Y signal to give a black and white and grey picture. The colour receiver also uses the Y signal to produce a black and white and grey picture.

Colour difference signals for grey scale

What happens to the colour difference signals when the same grey-scale scene is used? To answer this question, we must again perform a set of calculations. As we have already seen, the values of luminance reduce with darker shades of grey. If we now repeat the calculations to determine the values of the B − Y and R − Y signals, we find that they should all be zero. Table 12.4 shows that for any shade of grey, black or white, the B − Y signal is always zero. This is also true for the R − Y signal: it is also zero for any shade of grey, black or white in the scene.

For black or white or any shade of grey, the colour difference signals fall to zero.

This situation is ideal, because the encoder circuitry will not produce any spurious signals that may cause random noise on the receiver screen.

Table 12.4 The B − Y colour difference signals when viewing a grey-scale scene

	Blue camera output	minus	Luminance signal	=	B − Y
White	1.00		1.00		0.00
Very light grey	0.89		0.89		0.00
Light grey	0.70		0.70		0.00
Mid-grey	0.59		0.59		0.00
Grey	0.41		0.41		0.00
Dark grey	0.30		0.30		0.00
Very dark grey	0.11		0.11		0.00
Black	0.00		0.00		0.00

Weighting

What happens to the two chroma or colour difference signals as they go towards the main vision modulator at the transmitter? Refer back to Figure 12.1. The two colour difference signals are modulated onto their own carriers. These carriers are known as **subcarriers**. The two subcarriers are 90° out of phase with each other. They are then added together to produce a resultant subcarrier. This resultant subcarrier varies in amplitude and phase. Also added to the chroma signal is a special synchronising signal called the **colour burst**, which is needed in the receiver.

The resultant chroma signal and the burst signal are then added to the Y signal in another adder, and this resultant signal is fed to the main vision modulator at the transmitter. This is where the problems begin.

The main vision modulator produces a negatively modulated RF carrier signal. In the UK it will be at UHF. Remember from Chapter 10, Figure 10.6, that the resultant RF envelope's white level must not cause the carrier's amplitude to reduce to less than 18 per cent. If this did happen the RF carrier could cancel out altogether and cause problems in the receiver.

Highly saturated colours, such as those in the colour bar scene, could cause the above effect: that is, to overmodulate the RF carrier. Also, the colour bar signal might cause the RF carrier to exceed 100 per cent: that is, beyond the sync pulse tip. Figure 12.5a shows the effect caused by this overmodulation.

To prevent this, the amplitude of the two chrominance signals is reduced before they are added together and modulated onto the main RF carrier. This process is known as chroma signal weighting, or simply **weighting**. Figure 12.5b shows the result of weighting the two chrominance signals. The modulation level is now within the zero carrier and also the 100 per cent points.

The R − Y signal is reduced by about 10 per cent: more accurately, it emerges from an attenuator at 0.877 of its original size. The B − Y signal is reduced by about 50 per cent: the exact attenuator value is 0.493.

The original numerical values of the R − Y signal were produced by subtracting the Y signal from the red camera output. As we are altering its value by about 10 per cent, it would be incorrect to refer to it as R − Y. We must rename this colour difference signal, and the symbol that has been chosen is the letter V. Similarly, the B − Y signal is renamed U after the process of weighting.

Therefore

$$V = (R − Y) \times 0.877$$

and

$$U = (B − Y) \times 0.493$$

To find the values of the weighted chrominance signals we must perform a set of simple calculations. Table 12.5 shows all the workings for the weighted values of the R − Y and the B − Y signals.

Table 12.5 The final values for the weighted R − Y and B − Y signals

Colours	R − Y		Multiplying factor		V
White	0.00	×	0.877	=	0
Yellow	0.11	×	0.877	=	0.10
Cyan	−0.70	×	0.877	=	−0.61
Green	−0.59	×	0.877	=	−0.52
Magenta	0.59	×	0.877	=	0.52
Red	0.70	×	0.877	=	0.61
Blue	−0.11	×	0.877	=	−0.10
Black	0.00	×	0.877	=	0.00

Colours	B − Y		Multiplying factor		U
White	0.00	×	0.493	=	0.00
Yellow	−0.89	×	0.493	=	−0.44
Cyan	0.30	×	0.493	=	0.15
Green	−0.59	×	0.493	=	−0.30
Magenta	0.59	×	0.493	=	0.30
Red	−0.30	×	0.493	=	−0.15
Blue	0.89	×	0.493	=	0.44
Black	0.00	×	0.493	=	0.00

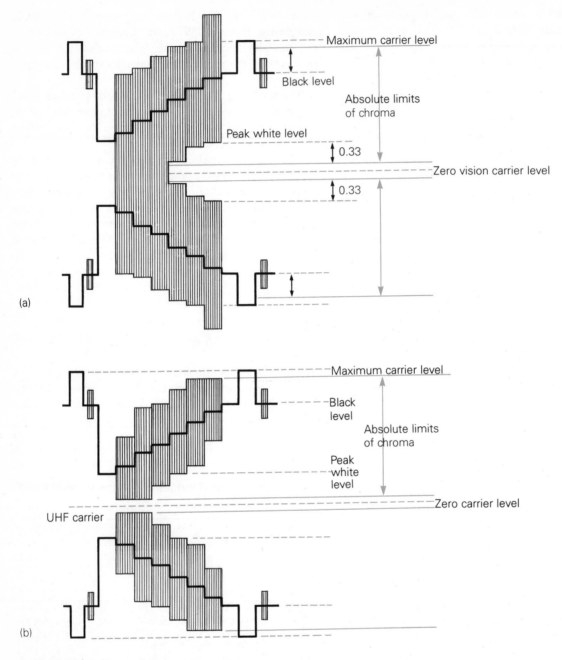

Figure 12.5 (a) Overmodulation caused by the chroma signals being too large; (b) the result of weighting the two chrominance signals.

Figure 12.6 shows the effect of weighting on the two chrominance signals. The results for V and U have been rounded up where necessary . In the receiver the two chrominance signals V and U must be restored to their original amplitude, otherwise incorrect colours would result. Remember that weighting is used only to overcome a problem that would have otherwise occurred during transmission.

Why are different amounts of weighting used? For the colour bar scene, the luminance value for yellow is 0.89 V. In the colour receiver the B − Y signal has to be able to 'turn off' the blue gun (as there is no blue in yellow). So if the luminance value is 0.89, then the B − Y signal must be −0.89. This represents a very large value of chrominance signal 'sitting' on the steps of the luminance signal. This causes the main vision carrier to cancel out: that is, go beyond 0 per cent.

Similarly, at the other end, for the blue bar, the red gun

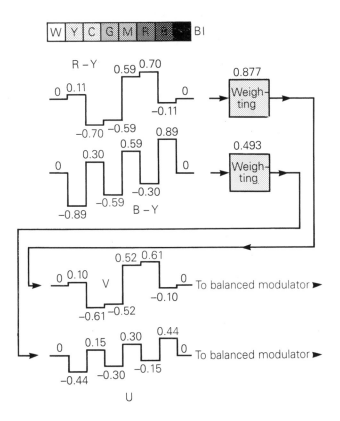

Figure 12.6 The effect of weighting on the two chrominance signals: R – Y in, V out; B – Y in, U out.

Balanced modulators

This section deals with modulation of the colour difference signals onto their own subcarriers. If you are still unsure of the principles of amplitude modulation, re-read the relevant section of Chapter 4. It is important to understand this fully to appreciate the signal processing at the receiver.

Figure 12.7b shows a 100 per cent amplitude-modulated

wave. One cycle of audio frequency is shown. We saw in Chapter 4 that such a wave is made up of three components, f_c, $f_c + f_m$, and $f_c - f_m$, where f_c is the carrier and f_m is the modulation signal. These three components are shown in Figure 12.7a. You can see that they are all related in a special way with reference to amplitude and phase. The side frequency $f_c + f_m$ is **lagging** the carrier (f_c) by 90°, and $f_c - f_m$ is **leading** the carrier. The amplitudes of the two side frequencies are half the value of f_c. Now, the amplitudes of the side frequencies depend on the amplitudes of the modulating signal. They are in fact produced when modulation takes place.

If the three components are added together graphically their sum is the amplitude-modulated wave shown in Figure 12.7a. So Figure 12.7a is made up of three special waves. The reverse is also true. Whether we know the values of the composite waveform or those of the separate components, we have the same information.

Now consider Figure 12.7c. This is the waveform that remains when the carrier is suppressed. You can obtain this waveform by adding together graphically the two side frequencies only.

The two side frequencies are produced by the modulator, but the carrier is not.

> A balanced modulator circuit produces side frequencies only, but not the carrier.

In this explanation, we shall use a single sine wave for the modulating signal. This results in two side frequencies. In practice, a whole range of frequencies would modulate the carrier. This would result in many sidebands being produced.

Figure 12.7c does not contain the carrier frequency component f_c, as it is not included in the addition of the waves, even though it may *appear* to be present. If there is no modulation, this waveform vanishes. (When there is no colour in a scene there are no colour difference signals, R – Y or B – Y, V or U. So when there is no colour, there is no subcarrier.) The subcarrier changes phase by 180° when the modulation signal changes polarity: that is, from positive to negative or from negative to positive.

> The balanced modulators produces no output when their inputs are at zero.
> The phase of the resultant carrier changes by 180° when the modulating signal changes polarity.

has to be turned off (ignoring the green gun for the moment). The blue gun must be turned on, so the B – Y signal must assist the luminance signal in this way. The luminance value for blue is 0.11. So then the B – Y signal must be at a value of 0.89 to make a value of 1 (0.11 + 0.89). Again, the value of the B – Y chrominance signal is still too large. When the B – Y chroma signal is added to the R – Y chroma signal at a later stage, overmodulation still occurs on the main vision carrier. The use of weighting overcomes the above problems.

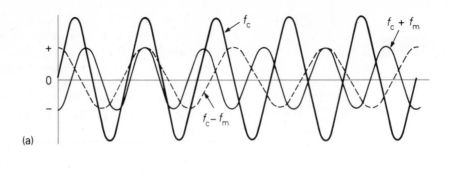

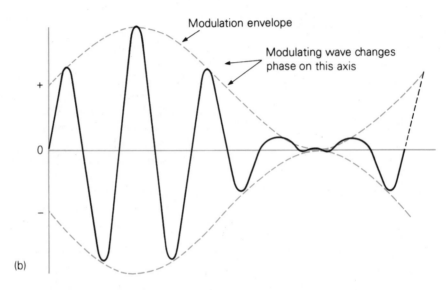

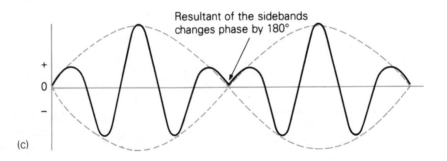

Figure 12.7 (a) Components of an amplitude-modulated wave; (b) sum of the three components; (c) sum of the sidebands only, i.e. no carrier (modulation product).

(Do not forget that the colour difference signals can change polarity. They can be positive or negative. This causes the subcarrier to change phase by 180°.)

At this point it is worth mentioning that a signal of this type requires the use of a special type of demodulator in the receiver. A normal type of AM demodulator will not work: the wrong signal output would result. With this type of modulation the carrier is effectively suppressed. The only way of demodulating a signal of this type is to 're-insert' the missing carrier. An oscillator circuit in the receiver acts as the missing carrier; its frequency is 4.43 MHz. The circuit is 'locked' in frequency and phase.

Figure 12.8a shows the input and output of the V balanced modulators, and Figure 12.8b shows the input and output of the U balanced modulator. The two balanced modulators shown in Figure 12.9 (which is part of the main block diagram in Figure 12.1) are fed from a common 4.43 MHz oscillator. Ignoring the inverter for

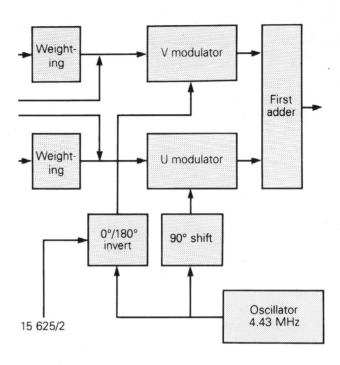

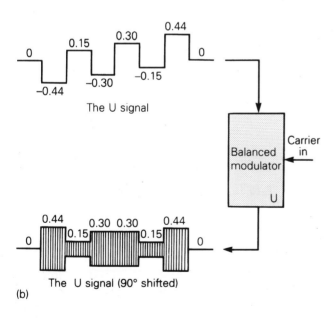

Figure 12.9 The V and U balanced modulators, taken from the main block diagram.

the moment, the V modulator would be directly fed, but the U modulator is fed via a 90° shift circuit. If both modulators were fed with subcarriers that were in phase with each other, their outputs would be in phase. Later on, when they were added together, they would lose their identity and would become inseparable. The 90° shift given to one of these subcarriers helps us to separate the composite chroma signal.

A 90° shift exists between the two subcarriers so that the composite chroma signal can be separated successfully in the receiver. Without this, separation would not be possible.

The first adder

In the first adder, the two separate subcarriers, V and U, are simply added together. The lower diagrams in Figure 12.8a and b show the peculiar shapes of these two envelopes. By definition, the modulated chrominance signals are $u = U \sin \omega t$ and $v = \pm V \cos \omega t$. However, for simplicity we shall continue to refer to them as U and V in this text.

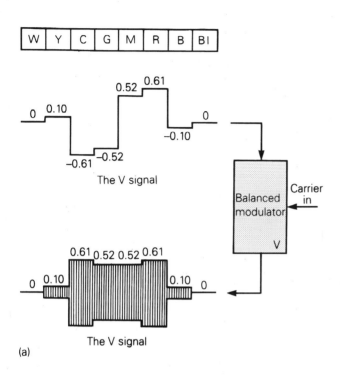

Figure 12.8 The input and outputs of the balanced modulators: (a) V; (b) U.

The modulated versions of the V signal and the U signal are added together in the first adder. The shape and size of the output will depend on the amplitude and polarity of the V signal, and the amplitude and polarity of the U signal.

At the V modulator output, the V subcarrier may be altering in amplitude and also changing in phase by 180°, depending on the original modulation. At the U modulator output, the U subcarrier may also be altering in amplitude and changing in phase by 180°. Both subcarrier outputs from the two modulators are also 90° apart from each other. This helps separation of the two original carriers in the receiver.

As the subcarrier outputs are sine wave shape, and are also 90° apart from each other, we can use phasors to represent them. The oscillator signal fed to the U balanced modulator is taken as the reference phasor with which all other phasors are compared. We can draw a horizontal line to represent the amplitude and polarity of the U signal (Figure 12.10a). Points to the right of O would indicate positive values of the U signal. The U chroma signal can change polarity: so points to the left of O would indicate negative values of U. This line is called −U, and is shown dashed.

Figure 12.10b shows the position of the phasor for the V subcarrier signal. Again, the original V chroma signal can change polarity. This would cause the V subcarrier to change phase by 180°. This is shown in a dashed line as the −V axis.

Superimposing Figure 12.10a onto Figure 12.10b gives Figure 12.10c. This complete phasor diagram indicates two chroma subcarriers, U and V, 90° out of phase with each other.

We can now determine the results of the two subcarriers added together. The values of the two chroma signals are shown in Table 12.6 for a set of colour bars. The first value for U would be −0.44 (for yellow). For V, it would be 0.1. These values are shown plotted on their respective axes in Figure 12.11. The resultant, R, will give the composite chroma value for the colour yellow.

To determine the amplitude (length) of this resultant, we have to use Pythagoras' rule. We cannot simply add the two figures together, because they are out of phase with each other. We would end up with the wrong answer. Figure 12.11 demonstrates the phasor addition:

$$\text{Resultant chroma} = \sqrt{V^2 + U^2}$$
$$= \sqrt{0.01 + 0.1936}$$
$$= \sqrt{0.2036}$$
$$R = 0.45$$

The resultant chroma (composite) signal is 0.45.

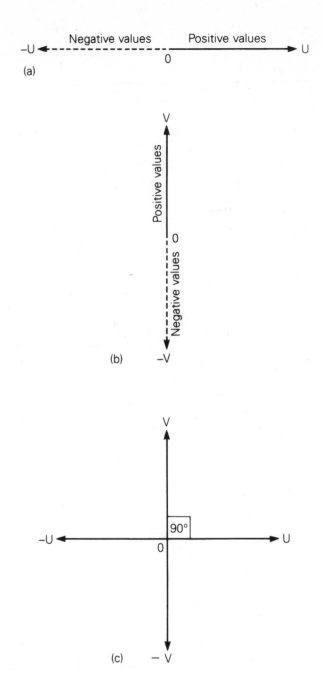

Figure 12.10 (a) Phasor for the U signal; (b) phasor for the V signal; (c) complete phasors for the V and U signals.

Table 12.6 Values of the two weighted chroma signals for the colour bar scene

	W	Y	C	G	M	R	B	Bl
U chroma	0.00	−0.44	0.15	−0.30	0.30	−0.15	0.44	0.00
V chroma	0.00	0.10	−0.61	−0.52	0.52	0.61	−0.10	0.00

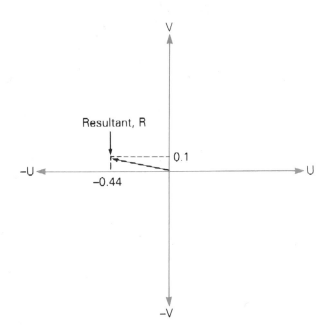

Figure 12.11 The resultant (R) chroma signal for the colour yellow.

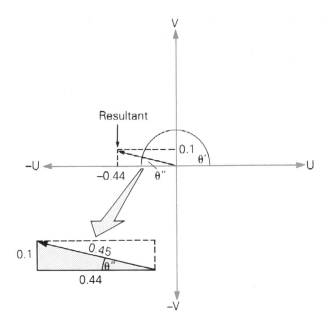

Figure 12.12 How to work out the angle by which the resultant leads U.

To work out the angle by which this resultant signal leads +U, we can use some basic trigonometry. If two sides of a right-angled triangle are known, then we can find the unknown angle. In this case we know all three sides. Let us use the tangent rule (Figure 12.12), which states:

$$\tan \theta = \frac{\text{opposite}}{\text{adjacent}}$$

First determine θ'' (we can ignore the fact that U is negative):

$$\tan \theta'' = \frac{0.10}{0.44}$$

$$= 0.23$$

then $\theta'' = \tan^{-1} 0.23$

$$= 12.95°$$

Now determine θ', given by $180° - \theta''$:

$$180° - \theta'' = 180 - 12.95$$

$$= 167.05°$$

The composite chroma signal will lead U by about 167°.

Figure 12.13 shows the resultant phasors (using the figures for the weighted chrominance signals) for the colour bars. The amplitudes and phase angles are shown. Numbers have been rounded up where necessary. This diagram is known as the **colour clock**. The positions of

the phasors indicate the colour that will be seen. For example, magenta is shown as having an amplitude of 0.59 and leading U by about 61°.

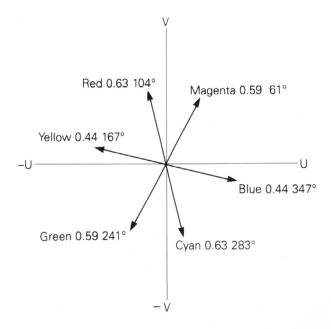

Figure 12.13 Positions of the colours for the colour bar signal. Phase angles and amplitudes are shown.

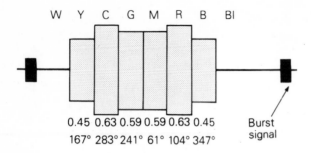

W Y C G M R B Bl

0.45 0.63 0.59 0.59 0.63 0.45   Burst
167° 283° 241° 61° 104° 347°   signal

Figure 12.14 The composite chroma signal with burst amplitudes and phase angles shown.

Figure 12.14 shows one line of composite chroma signal with all the amplitudes and phase angles.

The signal that emerges from the first adder is a subcarrier whose frequency is 4.43 MHz. The phase and amplitude will vary because of the polarity and size of the two original chrominance signals V and U. The resultant subcarrier will therefore be modulated in amplitude and phase. The correct name for this kind of modulation is **quadrature amplitude modulation** (QAM).

ACTIVITIES

1   For this activity you will need some graph paper (A4 size or similar), a simple calculator, and a protractor. Using the method that we have described to calculate the resultant chroma signal for weighted colours, construct a colour clock using the unweighted figures for the two colour difference signals B − Y and R − Y. You will find the results for these figures in Tables 12.1 and 12.2.

Draw the axes in the middle of the graph paper. Label the vertical axis R − Y and the horizontal axis B − Y. Do not forget the polarities (positive and negative: R − Y at the top, −(R − Y) at the bottom, B − Y on the right-hand side, −(B − Y) on the left. You need to choose a scale that will use up most of the paper. The highest number will be 1, so try 4 cm = 0.5. Keep the scale the same for both axes.

Using the results in Tables 12.2 and 12.1, plot the values on the axes. Start off with, say, R − Y = 0.11 and B − Y = −0.89. Continue in the same way and determine the resultant chroma signal amplitudes for the colour bar signal using Pythagoras' rule. Determine the angles, and check your results with a protractor. Label your diagram.

You will notice that the amplitudes and phase angles for the unweighted and weighted chroma signals are different from those for the weighted signals.

2   You can demonstrate the colour clock shown in Figure 12.13 by using a good general-purpose oscilloscope with an X–Y facility. You will also need a colour bar generator and the receiver service manual.

Adjust the colour receiver for a normal colour bar display. Locate the outputs of the V and U demodulators. At this point there should be the demodulated V′ and U signals, similar in shape to Figure 12.6. Check for this using the oscilloscope in the conventional way, with the timebase set to 10 μs/division. Connect channel 1 to the U signal and channel 2 to the V signal. Now, switch the input selector for each channel to the 'ground' or 'earth' position. The waveforms should vanish. Move the timebase control to the X–Y position. You should now see a spot on the screen. Use the shift controls to centralise the spot. Keep the attenuators set at the same value for each channel. When you have achieved this, move the input selectors for both channels to the d.c. position. You should now see on the oscilloscope something that resembles the colour clock shown in Figure 12.13. The important points should be the ends of each line. Now, reduce the colour control to zero. What happens to the display?

■ CHECK YOUR UNDERSTANDING

● The luminance signal is made up by taking the following proportions of the camera output signals: green, 59 per cent; red, 30 per cent; blue, 11 per cent. These proportions give the best overall result.

● The colour bar pattern is a special test pattern that is used to check the performance of the colour receiver. It consists of six colours (three primary and three complementary colours) plus white and black, making eight bars altogether.

● The relative luminosity curve is a graph that represents the frequency response of the average human eye.

● The luminance signal is given by Y = (0.59G) + (0.30R) + (0.11B)

● For colour, two extra signals are needed. These are known as chrominance signals. They consist of information about red and blue in the scene. For transmission, green chrominance signals are not needed, and are discarded.

● The red and blue chrominance signals give information about hue and saturation. The two chrominance signals are obtained by subtracting the Y signal from the red and blue camera outputs. At this point they are known as colour difference signals, R − Y and B − Y.

- White, black or any shade of grey in the scene will not produce any colour difference signals.
- The colour difference signals can have positive or negative values. These polarities must be preserved.
- The luminance signal is the information about scene brightness. It is used by both monochrome and colour television receivers.
- The chrominance subcarrier falls to zero for any black or white or any shade of grey in the scene.
- Weighting is needed to prevent the chrominance signals from overmodulating the main vision carrier.
- The $R - Y$ signal is reduced by about 10 per cent and the $B - Y$ signal is reduced by about 50 per cent.
- The weighted $R - Y$ signal and $B - Y$ signals are renamed V and U respectively.
- In the receiver, deweighting or increasing the amplitude of the V and the U signal is required. If this were not done incorrect colours would result.
- The two colour difference signals are modulated onto their own carriers. These carriers are called subcarriers. The frequency of these subcarriers is approximately 4.43 MHz.
- Suppressed carrier modulation is used; only sidebands are produced from the modulator. If the V and U signals are at zero, then no sidebands are produced from the modulators.
- If the V or the U signal changes polarity, then this causes its own subcarrier to change phase by 180°.
- Special types of demodulator are required for the V and the U signal in the receiver. Part of this process is the reinsertion of the suppressed carrier.

- The two carrier inputs to the V and U modulators are separated by 90°. This is done so that separation of the two can be achieved in the receiver.
- The modulated versions of the V and U signals are added together to produce a resultant chroma signal, which varies in amplitude and phase. This can be represented by using phasors.
- Pythagoras' rule can be used to determine the resultant length or amplitude of the chroma signal. The tangent rule can be used to obtain the angle by which the chroma signal leads the reference.
- The colour clock shows the resultant of adding together the V and U signals for the colour bar scene.
- Signals developed from the colour bar scene are used for clarity, instead of using chroma signals in a moving scene, which occur at random.
- The correct name for this type of modulation is quadrature amplitude modulation (QAM).

REVISION EXERCISES AND QUESTIONS

1 What is weighting?
2 The two chrominance signals are added together at the encoder in the transmitter. There is a 90° shift between the two reference subcarrier inputs. Why?
3 State the formula for deriving the luminance signal.
4 List the sequence of the standard colour bar pattern from left to right.

The PAL transmitter (continued)

The burst signal

In order to demodulate a signal that does not have any carrier (because it is suppressed in the transmitter encoder), it is necessary for the colour receiver to generate one: that is, an oscillator signal that is at an identical frequency and phase. The frequency chosen for this oscillator is approximately 4.43 MHz or, more accurately, 4.433 618 75 MHz.

In order for the oscillator to run at this exact frequency and phase, the transmitter needs to send out with the composite chroma signal a sample of the original 4.43 MHz oscillator. In fact one sample is sent out after every line. This sample is called the **colour burst** signal or simply the **burst** signal.

The main job of the colour burst is to act as a timing signal for the reference oscillator in the receiver. However, there are two other reasons why the burst signal is needed. First, it acts as an identification signal for the inverted V line. This will be explained later when we deal with the PAL system. Second, it can act as an indication of the strength of the chroma signal: that is, it can be used for **automatic chrominance control** (ACC).

The burst signal must be built in as part of the composite video signal, but it must not interfere with the rest of the chroma signal or video signal. How much of it do we send out? The word 'burst' would indicate that only a small amount of signal is required. In the PAL system about ten cycles are transmitted.

Where do we put this burst signal? It is slotted onto the back porch region, which immediately follows the line sync pulse. There is no video signal in this area. Figure 13.1 shows the front porch and line sync pulse. The burst signal is shown in place on the back porch.

The back porch occupies a time duration of about 5.8 μs. Ten cycles of burst signal occupy a time duration of 10 $\times$ the time duration of one cycle of burst. = 10 $\times$ 0.225 μs = 2.25 μs. The burst peak-to-peak value should be at the same height as the line sync pulse.

The burst signal will be discussed in greater detail a little later on.

The PAL system

Up to this point, the description that has been given is common to both the American NTSC and the British PAL colour television systems. The only minor differences are that the choice of frequencies is slightly different. For example, the chrominance subcarrier frequency for PAL is 4.43 MHz instead of 3.58 MHz for NTSC. The highest video frequency for PAL is 5.5 MHz instead of 4.5 MHz. Also, the number of lines used for the PAL system is 625 compared with 525 lines for the NTSC system. NTSC also uses a 60 Hz field frequency.

In the NTSC system, an unwanted phase change in the chrominance signal would produce incorrect colours on the television screen. Any analogue signal, whether complex or of simple sine wave shape, can undergo changes of phase, especially when passing through, say, a number of amplifiers. In straightforward amplifier systems such as those used in audio, a phase change of the signal is not too important, as the phase of the signal doesn't have much effect on the music quality. In colour television signals, the subcarrier is the chroma signal, and it uses phase and amplitude modulation (QAM). The phase determines the colour, and the amplitude determines the saturation level. A common emitter transistor amplifier, such as the chroma amplifier in the receiver, can introduce unwanted phase changes in the chrominance signal and cause the above effects. This effect is known as **differential phase distortion**.

NTSC receivers have an extra manual adjustment, which can be adjusted to allow for the effects of this distortion. It is known as the **tint control**, and it is

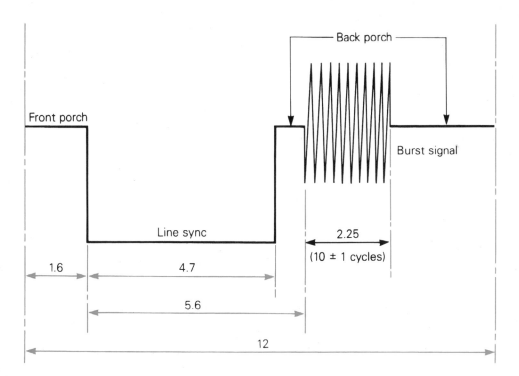

Figure 13.1 The burst signal sitting on the back porch (all times in μs).

adjusted for best overall colour. In modern PAL receivers, adjustment to remove these phase errors is automatic, eliminating the need for this extra control. It is automatic, in the sense that these phase errors are compensated for. How does this work? PAL stands for 'phase alternation line'. This suggests that something is done to the chroma signal on a line-by-line basis.

Imagine that the studio camera is pointing at a scene that is pure magenta. What would be the amplitude and phase of the resultant chroma signal? A quick check back with the colour clock in Figure 12.13 will show that the amplitude is 0.59 with a phase that is leading U by about 61°. Figure 13.2a shows the resultant phasor for this chroma signal.

Now, what would happen if this chroma signal were to suffer from a change in phase? This could occur in the receiver itself, possibly caused by mistuning. Figure 13.2b shows the resultant phasor for this chroma signal, but this time with a phase lag of 10°. Clearly this would cause a more bluish magenta on the receiver screen. The reason is that in the receiver the U demodulator would produce more signal output than the V demodulator. Therefore a phase lag or lead results in incorrect colours on the receiver screen.

The PAL system was designed to overcome this effect. In the PAL system the V subcarrier is inverted at the transmitter every second line. For the next chroma line following chroma line, the inversion takes place again. is sent out as normal: that is, not inverted. For the

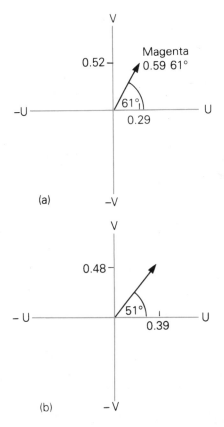

Figure 13.2 (a) The resultant phasor for the colour magenta. (b) The resultant phasor, with a 10° lag. When demodulated in the receiver, U will be equal to 0.39 instead of 0.29. V will be 0.48 instead of 0.52. The result will be a bluish magenta.

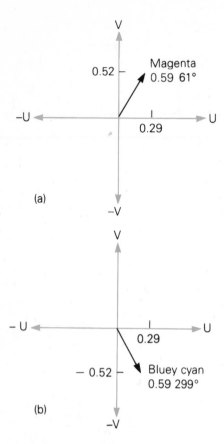

Figure 13.3 (a) Chroma line *n*; (b) chroma line *n* + 1.

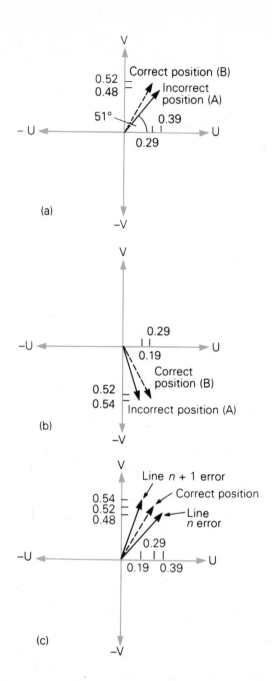

Figure 13.4 (a) Chroma line *n* and (b) chroma line *n* + 1, showing a phase lag after transmission. Dotted line: transmitted as the correct phasor. Solid line: received as the incorrect phasor. In (b), A is still lagging B by about 10°. (c) Two lines of chroma signal; line *n* error produces a bluish magenta; line *n* + 1 error produces a reddish magenta.

The signal inversion that takes place at the transmitter is corrected again in the receiver. The U reference subcarrier is not affected.

Now look at Figure 13.3a. This represents pure magenta: 0.52 of V and 0.29 of U. This is the correct colour. We shall call this chroma line *n*. It is the same as the diagram in Figure 13.2a. On the next chroma line (line *n* + 1) this would be transmitted as 0.52 of −V and 0.29 of U (Figure 13.3b). This would produce a line of bluish cyan. The receiver corrects for this line by reinverting the V subcarrier so that the colour magenta is produced, as line *n*. The receiver must correct the inverted line of chroma, otherwise the wrong colour(s) will be produced on the screen for that particular line.

Now, what happens when the transmitted chroma signal has suffered from a phase lag? Again, we can use a figure of 10°. See Figure 13.4a for line *n*. On chroma line *n* + 1 (Figure 13.4b) the received chroma signal is still showing a 10° lag, but this time in the opposite direction. A is still lagging B. The receiver corrects for line *n* + 1.

Figure 13.4c shows two lines of chroma with line *n* error and line *n* +1 error. After the V signal is inverted, the 10°

lag becomes a 10° lead. The 10° lead (line *n* + 1) averages out with the 10° lag (line *n*). Line *n* produces a bluish magenta and line *n* + 1 produces a reddish magenta. At normal viewing distances these two different coloured lines can be **optically averaged**: that is, the eye can mix

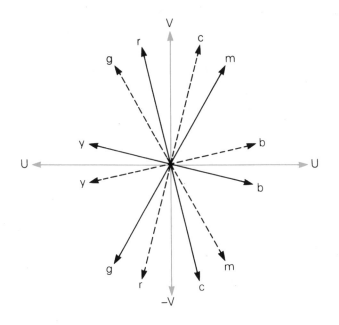

Figure 13.5 The colour clock for the NTSC chroma line (solid) and the PAL chroma line (dotted).

the two slightly different colours and as a result see the correct colour. In practice, the use of interlaced scanning means that two lines of each incorrect colour would be seen.

Let us take a look at the effect that the PAL system has on the colour bar signal colour clock. Figure 13.5 shows in dotted lines the new positions for the composite chroma signal. This is brought about by V signal inversion.

What causes the V chroma signal to invert or change its polarity by 180°? The answer is the **inverter**, shown in the main block diagram in Figure 12.1. The 0/180° inverter is connected in series with the oscillator feed to the V balanced modulator. The 0/180° switch, or **vertical axis switch** as it is sometimes known, is responsible for

reversing the phase of the oscillator signal that feeds into the balanced modulator.

Figure 13.6 shows two successive lines of V at the output of the balanced modulator. The individual cycles of subcarrier cannot normally be seen because they are too close together. In this diagram we have tried to show what they would look like if it were possible see them.

The process of error cancellation and separation will be looked at in Chapter 20.

PAL D

Early PAL colour television receivers relied on the principle of optical averaging. They were called **simple PAL receivers**. In practice, phase errors of more than 5° would probably be noticeable at normal viewing distances as something known as Hanover bars. This is why the **PAL delay line** was introduced.

A **PAL D** receiver is far superior to a simple PAL receiver, because it **electrically averages** the phase errors to give the correct colour on the screen. It does this by storing or delaying one line of chroma signal.

Let us assume that line *n* chroma is being received. This line is stored for 64 μs. When line *n* + 1 chroma is received, both the stored and the direct lines of chroma are fed to a matrix circuit, where they are averaged out to produce the correct colour. The electronic store (called a delay line) and matrix circuit also help to average out when there is a change in chroma signal from one line to the next, as in a picture of horizontal bars of different colours, for example. The delay line and matrix circuit in the receiver not only get rid of phase errors, they can also separate the V and the U signals. This will be discussed later on when we look at the colour receiver.

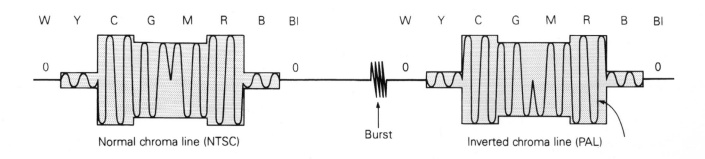

Figure 13.6 Note the inverted subcarrier within the PAL envelope.

The swinging burst

We mentioned earlier that the colour receiver must re-invert the inverted V chroma signal so that the correct colour may be reproduced. For this to happen, the receiver must be told when to perform this operation. The burst signal, as well as controlling the receiver reference oscillator, is also responsible for telling the colour receiver when the chroma subcarrier has been inverted. The frequency of the burst is 4.43 MHz.

The phase of the burst is different for the normal V chroma line and the −V chroma line. Figure 13.7 shows the two phases of the burst signal for two lines of chroma, n and $n + 1$. The burst is shown only in relation to where the V subcarrier and the U subcarrier would be. It is not part of the composite chroma signal because it is transmitted at a different time.

The colour receiver 'knows' the phase of the burst signal, and acts on it by reversing the phase of the V reference signal. The fact that the burst is swinging in phase does not affect its main job of 'locking' the receiver reference oscillator. In fact the average phase of the burst signal is along the −U axis. The receiver reference oscillator is locked at 90° to this average phase: that is, along the V axis. The burst swings through 90°, from an angle of 135° on chroma line n, to 225° for chroma line $n + 1$.

The burst signal does not interfere with the chroma signal because the burst is on the back porch and does not appear in the same 'time domain' as the chroma signal or even the video signal.

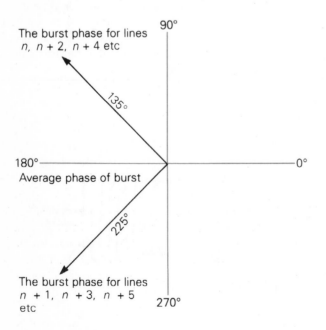

Figure 13.7 The swinging burst.

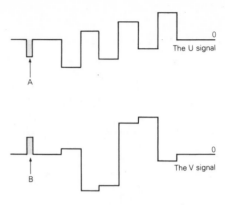

Figure 13.8 The keying pulses used to produce the burst signal. Pulse A produces the burst signal shown in Figure 13.9; produces the burst signal shown in Figure 13.10.

The burst signal is produced at the transmitter by feeding into the balanced modulators suitably shaped pulses known as **keying pulses**. Pulse A in Figure 13.8 is negative going, and will produce a sine wave output at 4.43 MHz from the U balanced modulator. The phase will be 180° referenced to the U axis (Figure 13.9). Pulse B in Figure 13.8 is positive going, and again will produce a sine wave output at 4.43 MHz from the V balanced modulator. The phase will be 90°, again referenced to the U axis (Figure 13.10).

The two independent burst signals, which are 90° apart from each other, are added together in the first adder (Figure 12.1) together with the two chrominance signals. The resultant burst is shown in Figure 13.11. The phase is shown as being 135°. This will be the phase for all NTSC types of chroma line. For PAL, the burst signal changes phase, and this is brought about by the inverter, shown in Figure 12.1. This not only inverts the V chroma signal but also inverts the burst emerging from the V modulator (Figure 13.12).

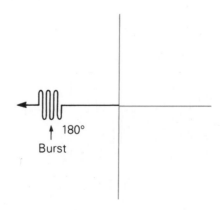

Figure 13.9 The phase of the burst signal at the output of the U modulator.

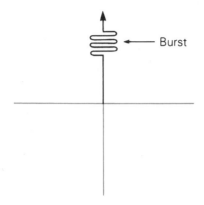

Figure 13.10 The phase of the burst signal at the output of the modulator for NTSC lines.

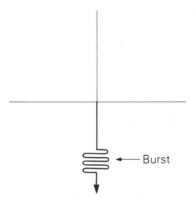

Figure 13.12 The burst signal at the output of the V modulator on PAL lines.

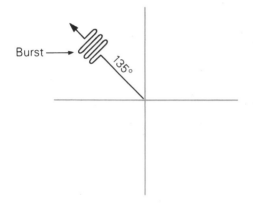

Figure 13.11 The resultant burst at the output of the first adder for all NTSC lines.

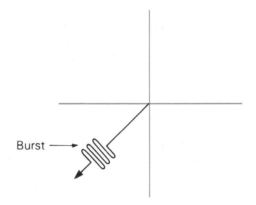

Figure 13.13 The result of adding the burst from Figure 13.12 to the burst signal in Figure 13.9 for PAL lines $n + 1$, $n + 3$, $n + 5$ etc. Output of first adder.

Figure 13.13 shows the result of adding the two individual burst signals together in the first adder for all the PAL colour lines. On normal chroma lines the burst will be 135°, and for PAL chroma lines the burst will be 225°, with reference to the U axis. The keying pulses must be carefully positioned and must also be of the correct amplitude, otherwise the phase and size of the burst signal will be wrong.

> The burst is swinging in phase. On normal colour lines it is at 135°. On PAL colour lines it is 225°. The functions of the burst so far are: to synchronise the receiver reference oscillator; to tell the colour receiver that the chroma line is either an NTSC type or a PAL type.

The subcarrier frequency

It is essential that the chrominance signal be transmitted within the existing bandwidth occupied by the monochrome signal. Originally, it was thought that because extra information was needed for the colour signals, extra bandwidth would be needed.

The colour subcarrier contains information about the colour content in the scene. The word 'subcarrier' indicates that it is a carrier within a main carrier. The main carrier is the main vision carrier that is transmitted from the transmitting aerial. The colour subcarrier, along with the luminance signal, modulates the main vision carrier. This produces an amplitude-modulated envelope (see Figure 12.5).

As you now know, the subcarrier frequency is 4.43 MHz. The choice of frequency is important: if it had been poorly chosen it would have caused a large amount of interference on the receiver screen. This interference is known as **dot pattern interference** or **dot crawl**, and it is just visible on a monochrome television receiver. Remember that the colour subcarrier is added to the luminance signal in the second adder, and this is why interference may be seen on a monochrome receiver, where the subcarrier together with the luminance signal modulates the electron beam.

In a colour receiver the subcarrier can be 'tuned out' or rejected, but in a monochrome receiver rejecting the subcarrier will reduce the definition of the display picture at that chosen frequency. Most manufacturers do not usually bother with rejection of the colour subcarrier in a monochrome receiver, because of the effect that it will have on picture definition. The expense of using another tuned circuit was also a deciding factor. Therefore, it was always important to get the choice of subcarrier frequency right so that this unavoidable interference would be at a minimum.

If any television picture is examined it is noticeable that the average scene content is of low definition: that is, video signals from very low frequencies up to about 2 MHz. High video signal frequencies exist mainly in test signals and certain fine patterns.

If the subcarrier had a frequency of 1.5 MHz a coarse dot pattern would appear along each line of the received picture. This pattern would be similar in size to the frequency grating bars shown in test card F (Appendix 2). This would be particularly visible on monochrome receivers, as usually no effort is made to reject the subcarrier. What is worse is that the dots will combine line by line to form an overall pattern on the picture, depending on the phase of the subcarrier. To make such a dot pattern negligibly fine, the subcarrier must have a

frequency that is as high as possible in the video range.

The chrominance signal needs a bandwidth of about 2 MHz. Placing the chrominance signal higher up towards the 4.5 MHz position would allow enough room for the upper and lower sidebands. The frequency is approximately 4.5 ± 1 MHz.

Two Americans, Gray and Hefele, discovered many years ago that the RF spectrum produced by a television signal contained a number of gaps, or areas where there was no energy. These gaps therefore could be used for other signals.

Figure 13.14 attempts to show this effect. Bundles of energy are shown, which are due to the repetitive nature of the video signal with its line and field sync frequencies. The spacing of these bundles is at 15 625 Hz. The same reasoning applies to the subcarrier spectrum.

Figure 13.15 shows an enlarged view of part of the RF spectrum. The dominant modulation frequency is the line repetition frequency, 15 625 Hz. The waveform is complex. It is therefore equivalent to a fundamental sine wave of 15 625 Hz plus all the harmonics: $f_1 + 2f_1 + 3f_1 + \dots$ etc. As the video bandwidth is 5.5 MHz, TV system I allows up to h_n harmonics, where h_n is given by

$$h_n = \frac{\text{Video bandwidth}}{\text{Line frequency}}$$
$$= \frac{5.5\,\text{MHz}}{15\,625\,\text{MHz}}$$
$$= 352 \text{ harmonics}$$

There are other repetition frequencies in the RF signals, such as picture frequency (25) and field frequency (50), but in general these signals consist of narrow bands of energy at the places indicated. There are very large gaps in the RF

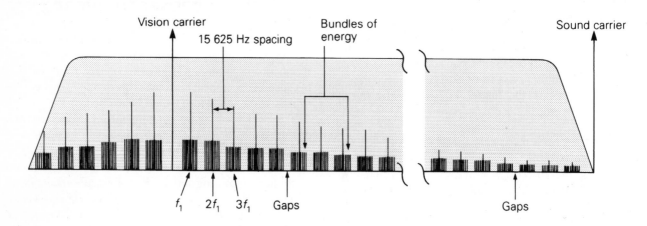

Figure 13.14 RF spectrum showing gaps produced by the modulation (not to scale). There would in fact be many more bundles of energy than shown here.

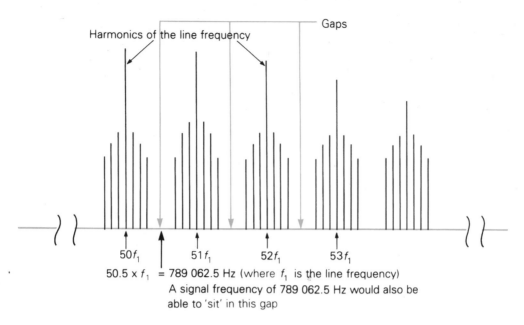

Figure 13.15 An enlarged view of a section of the RF spectrum.

spectrum at line frequency. If the subcarrier components are to interlace exactly within this RF spectrum, then the subcarrier frequency must be an odd multiple of one half of the line frequency.

Consider Figure 13.15. One half of line frequency, $f_1/2 = 15\,625/2 = 7812.5$ Hz. All **even** multiples of 7812.5 fall on harmonics of the line frequency: for example $100 \times 7812.5 = 781\,250$ Hz, which is 50 times the line frequency. Similarly 102×7812.5 Hz $= 51 \times f_1$. All **odd** multiples of 7812.5 fall exactly between harmonics of the line frequency: for example, $101 \times 7812.5 = 789\,062.5$ Hz. This is $50.5 \times f_1$. $103 \times 7812.5 = 51.5 \times f_1$... and so on. It is clear

that odd multiple $\times$ 1/2 line frequency, say $101 \times f_1/2$, can be written as $50.5 \times f_1$.

For example, in Figure 13.15, in order for an extra signal to be fitted into the gaps shown, then we may choose a frequency that is $50.5 \times f_1$. This frequency works out to $50.5 \times 15\,625$ Hz $= 789\,062.5$ Hz. This signal would fit into the middle of the gap shown without causing too much interference.

Figure 13.16 shows the approximate position of the subcarrier and its sidebands in the RF spectrum. It is sitting between the 283rd harmonic of the line frequency (4 421 870.00 Hz) and the 284th (4 437 500.00 Hz). In order

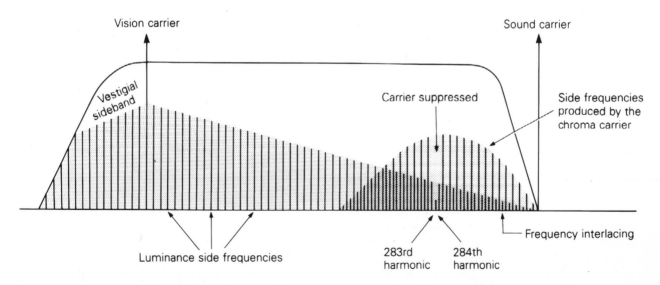

Figure 13.16 The position of the chrominance subcarrier within the RF spectrum (not to scale).

for the chroma signal to sit in the middle of these two points then the frequency should be 283.5 times the line frequency. (Alternatively $567 \times 15\,625/2$ satisfies the definition.) This gives $283.5 \times 15\,625$ Hz $= 4\,429\,687.5$ Hz or 4.429 687 5 MHz.

If the NTSC system had been adopted in the UK this probably would have been the chosen frequency for the subcarrier. However, because the V signal alternates in the PAL system, and the pattern obtained on the screen of a monochrome receiver reinforces itself on a line-by-line basis, some alterations are needed.

One complete picture line and sync pulse with front and back porch is equivalent to 283.5 cycles of subcarrier. (The period of one cycle of subcarrier multiplied by 283.5 $= 64\,\mu s$.) This half-line offset (the odd half) is changed to 3/4 line or quarter-line offset for the PAL system. This modification minimises the dot pattern effect due to the switching of the V signal. To reduce the effect of patterning still further, 25 Hz is added to the final figure.

The final figure then for the subcarrier frequency f_{sc} used in the PAL system is now given by

$$f_{sc} = (283.75 \times 15\,625) + 25 \text{ Hz}$$
$$= 4\,433\,618.75 \text{ Hz}$$
$$f_{sc} = 4.433\,618\,75 \text{ MHz}$$

It is neither necessary nor practical to attempt to draw the various screen patterns that would be created by the subcarrier signal on the luminance signal. The frequency chosen for the subcarrier signals gives the minimum interference effect upon the monochrome receiver.

This method of 'slotting in' extra information is known as **frequency interleaving** or **frequency interlacing**.

Cross colour

It is possible for the luminance signal to interfere with the chrominance signal. This happens with high-frequency luminance signals, which correspond to about 4.4 MHz definition. This causes havoc in the receiver decoder. The colour decoder in the receiver 'thinks' that the high-frequency luminance signal is the chroma signal, and tries to decode it. The effect produced on the screen consists of coloured patterning in the area of this luminance. This is known as **cross colour**.

In a monochrome receiver, the subcarrier can produce an effect called **cross luminance** if the scene in front of the camera consists of a saturated colour. Let us assume that the colour in question is green. The luminance value for this should be about 0.59 (assuming a perfect system). In a monochrome receiver, very few manufacturers

bother to suppress the subcarrier. So the luminance signal appearing at the cathode of the CRT will have on it the chrominance subcarrier, which for a saturated colour could be of high amplitude. Effectively, there is a 4.43 MHz sine wave sitting on the top of the luminance signal. The characteristic curve of the picture tube could rectify this subcarrier, as a result of which the picture brightness could increase, giving an incorrect shade of grey.

The luminance delay line and second adder

The luminance delay line (see main block diagram in Figure 12.1) slows down the luminance signal so that it arrives at the second adder at the same time as the composite chroma signal. Wide-band signals (0–5.5 MHz) travel faster through circuitry than narrow-band signals, and the chrominance signal is a narrow-band signal. A typical delay introduced would be around about 0.6–1 μs.

The second adder is where the composite chrominance signal and the luminance signal meet and join together to produce the composite video waveform. After the second adder, sync pulses (line and field) are added to produce the full signal. Figure 13.17 shows one line of this signal.

The signal shown in Figure 13.17 is fed to the main vision modulator at the transmitter. It is then radiated in the form of an amplitude-modulated RF carrier wave signal similar to Figure 12.5. The sound signal is radiated as an FM carrier wave, which is 6 MHz (or, more accurately, 5.996 MHz) in frequency higher than the vision carrier.

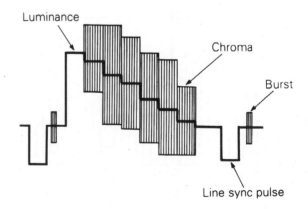

Figure 13.17 One complete line of composite video signal (luminance and chrominance).

CHECK YOUR UNDERSTANDING

● The colour burst signal is transmitted along with the chroma signal. The burst signal 'sits' on the back porch. Its frequency is 4.43 MHz.

● One use of the burst is to synchronise the receiver reference oscillator. Another use is to identify the inverted V signal.

● Ten cycles of burst are transmitted. It occupies a time duration of about 2.25 μs.

● For NTSC, any unwanted phase change in the chroma signal would produce different colours on the screen of the CRT.

● For PAL, the V signal is inverted on every other line. On the chroma line following, it is reverted back. The U signal is left alone. Any unwanted phase lag or lead will be compensated for by reversing the error: that is, a lag will become a lead or a lead will become a lag.

● The inverter circuit in the main block diagram causes the inversion of the V chroma signal. It is known as the vertical axis or PAL switch.

● If a simple PAL receiver is used, optical averaging takes place. For a delay line PAL receiver, electrical averaging is used.

● The burst signal swings in phase. On the normal chroma lines 1, 3, 5, 7, ..., it is at 135°. On PAL chroma lines 2, 4, 6, 8, ..., it is at 225°. The average phase of the burst signal is along the U axis, but it swings through 90° for every line from 135° to 225°.

● Keying pulses are used to produce the burst signals. They are fed into the balanced modulators along with the V and the U signals.

● The choice of frequency for the subcarrier is important. It is inevitable that some interference will take place, and show itself as dot crawl on a monochrome receiver picture. To make the dot pattern as fine as possible the subcarrier frequency chosen must be as high as possible.

● Both sidebands of the chrominance signal must also be allowed to fit into the available space within the luminance sidebands.

● Quarter, or three-quarter line offset for the subcarrier is used. This means that the frequency chosen will place the subcarrier 'over to one side' (over towards the right) and not directly in the middle of the gap. Therefore the frequency of the subcarrier is equal to

$$f_{sc} = f_l\left(284 - \tfrac{1}{4}\right) + 25$$
$$= f_l(283.75) + 25$$
$$= 4.433\,618\,75 \text{ MHz}$$

where f_l is the line frequency, 15 625 Hz.

● Certain frequencies of the luminance signal can interfere with the chrominance decoder in the receiver. This effect is known as cross colour.

● Delay lines are used to slow down the luminance signal so that it may arrive at the same time as the chrominance signal. For the composite video signal, AM is used. The vision carrier is transmitted in the UHF band of frequencies.

● The sound carrier uses FM. It is transmitted at 6 MHz above the vision carrier.

REVISION EXERCISES AND QUESTIONS

1 The burst signal is transmitted along with the chrominance and the luminance signal. State one function of this burst signal.

2 With reference to colour television transmissions, what do you understand by the terms 'frequency interleaving' or 'frequency interlacing'?

3 In PAL, the phase of the V chrominance signal is reversed every other line. On line n it is 90°, on line $n+1$ it is 270°, on line $n+2$ it is 90°, and so on. Why is this done?

4 The frequency of the colour subcarrier is specified as 4.433 618 75 MHz ±1 Hz. Why does it have to be so exact?

The monochrome receiver

Introduction

Having looked at monochrome and colour television transmission, we now look at receivers. This chapter is the first of six that look in detail at the monochrome receiver.

We begin in this chapter by taking a brief look at the UHF tuner unit and IF stages, before explaining the vision demodulator and video amplifier. In Chapter 15 we look at the synchronising separator, and in Chapter 16 at the field timebase. Chapter 17 explains the functions of the line output stage and EHT of a typical monochrome receiver. Chapter 18 looks at the sound IF channel, and Chapter 19 deals with the monochrome CRT.

The UHF tuner unit

The function of the tuner unit in any television receiver, whether monochrome or colour, is to select, amplify and change the incoming RF signals. By 'change' we mean to alter the original carrier frequencies into frequencies that are easier to handle. These different frequencies are known as **intermediate frequencies** (IF).

The tuner unit must be selective: that is, it must be able to accept the wanted signals and reject the unwanted ones. It must also provide sufficient gain to produce a reasonably high signal-to-noise ratio.

The selected carrier frequencies are changed into the required intermediate frequencies by the mixer/oscillator stage. It does this by using the principle of **heterodyning** (mixing) (see Chapter 5).

Some form of isolation from the aerial must also be provided. For mains-operated receivers it is possible that the aerial system might become live because of the direct connection to the television receiver chassis or metal framework.

The gain of the tuner unit must be made variable so that the receiver can operate in strong or weak signal areas.

AFC (automatic frequency control) may be provided to prevent tuning drift. This is essential in all colour

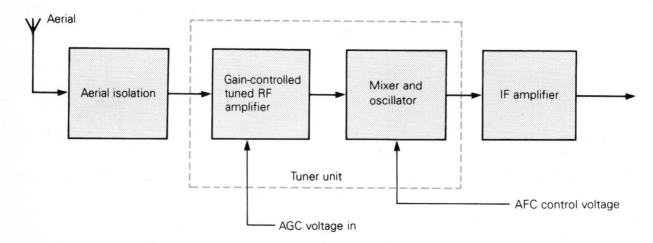

Figure 14.1 Block diagram of the tuner unit and first IF stage.

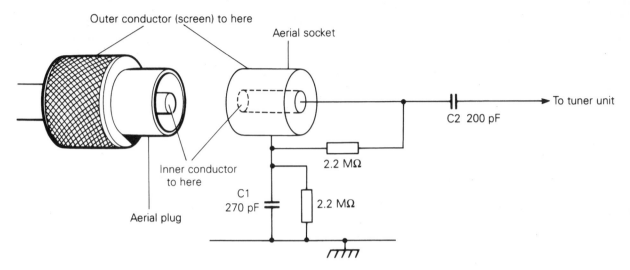

Figure 14.2 Circuit of a typical aerial isolation unit.

receivers, but it is optional in some small monochrome portable receivers.

Figure 14.1 shows a block diagram of a complete tuner unit. Also shown are the aerial isolation block and first IF amplifier.

Figure 14.2 shows an aerial isolation circuit. Capacitors C1 and C2 have a low value. They offer a very small reactance to radio frequencies but a very high reactance to mains frequencies (50–60 Hz).

Remember that

$$X_c = \frac{1}{2\pi f\, C}$$

This fact effectively provides electrical isolation for the

whole of the aerial circuit from the metal chassis of the receiver. The build-up of static charges on the aerial system is prevented by the high-value resistors. Any static charges will discharge through these resistors to the chassis. The aerial isolation unit is a vital safety component, and if a fault should become apparent – for example, when the socket is damaged – then the complete unit must be replaced.

The tuner unit is completely self-contained (Figure 14.3). We do not recommend servicing or adjusting the tuner unit, as specialised test equipment is usually needed. A detailed explanation of the workings of the tuner unit is outside the scope of this book. However, a brief description will be given with reference to

Figure 14.3 Typical tuner unit.

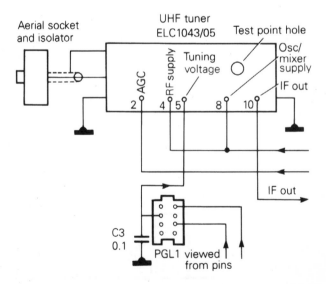

Figure 14.4 A receiver tuner unit (Ferguson 1600 series).

Figure 14.4, which shows the tuner unit as a simple block. This is the way most manufacturers now show this component. The circuitry within the block is normally as follows.

An RF amplifier transistor is usually connected in the common base configuration. This mode of operation is ideal for matching purposes: that is, low input impedance and high output impedance. The stage is 'tuned': it is selective, and it is also gain controlled (as mentioned earlier). The tuning of this stage is carried out by using varicap diodes connected as part of the tuned circuits within (see Chapter 5). As the RF amplifier is connected in the common base mode, it also acts as a **reverse buffer**: it prevents any signals from the oscillator from being fed back via the aerial cable to the aerial, and hence re-radiating and causing interference to other receivers.

The different channels are selected by a **tuning voltage** connected to pin 5 of the unit (more on this later). Pin 2 is an AGC control voltage, which is normally delayed or held off until a very strong RF carrier signal is received. This means that the RF amplifier runs close to or very nearly at maximum gain; the gain reduces only when a very strong signal is received.

The mixer/oscillator

The job of the mixer stage is to combine the incoming carrier signals with the signal produced from the oscillator. This function is usually carried out by one transistor operating in the common base mode.

The transistor may be operating as a self-oscillating mixer, working at a frequency that is 39.5 MHz above the vision carrier. The carrier signals and the oscillator signal may be mixed together in a tuned circuit or load. The tuning of the oscillator is controlled by the same variable voltage, which is supplied from pin 5 of the unit. Varicap or varactor diodes are employed. As for the RF amplifier, these diodes are connected as part of the tuned circuit, which (for the oscillator) alters its frequency. The diodes are reverse biased: their internal capacitance reduces in value for an increase in tuning voltage. The coverage of the oscillator would have to be approximately 490–900 MHz for bands IV and V. This will enable channels from 470 MHz to 854 MHz to be received.

Varicap tuners have been in use for over 20 years. They have replaced the old ganged capacitor type. They are cheaper to make, are reliable, and have no moving parts. They are more suited to receivers that have touch-tuning facilities, and also remote control. The switching unit for the channel-change functions can be put into a convenient position inside the receiver cabinet; it does not carry any RF signals, just a variable d.c. It does not need to be physically close to the tuner unit itself.

Figure 14.5 shows the circuit diagram of a typical channel selector unit. The voltage for the tuning potentiometers R401–R407 is fed in on pin 2 of SKT1. This voltage is obtained from pin 2 of PLG1 (Figure 14.4) via R3. Channel number 7 is shown to be selected. R407 is switched into circuit. The voltage at pin 3 of SKT1 will now depend on the position of the slider on R407. When another button is selected, say number 6, channel 7 will be disengaged (pop out), and the new channel will be selected. The sliders on each of the potentiometers will normally be set to different positions, giving different values of voltage for pin 5 of the tuner unit. The voltage

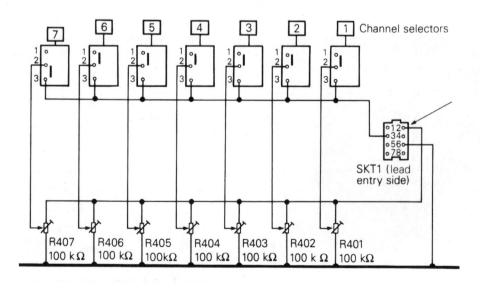

Figure 14.5 A channel selector unit (Ferguson).

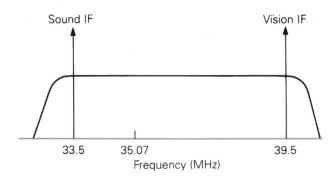

Figure 14.6 The RF spectrum at the output of the tuner unit.

range for this type of tuner is normally 0–33 V.

Pins 4 and pins 8 supply the power (12 V line) to the RF amplifier and the mixer/oscillator respectively.

Pin 10 on the tuner (Figure 14.4) is the IF output. There are two signals at this point: the vision IF at 39.5 MHz and the sound IF at 33.5 MHz. The vision IF signal also carries the chrominance signal. This will generate its own IF at 35.07 MHz (39.5 MHz – 4.43 MHz = 35.07 MHz).

Figure 14.6 shows the RF spectrum at the output of the tuner (pin 10). Note that the vision signal is now higher in frequency than the sound signal. This is because of the mixing action of the two signals within the tuner.

For example, let us say that the TV receiver is tuned into the UK channel BBC 2 from the Winter Hill transmitter. The vision carrier is radiated at a frequency of 799.25 MHz. The sound carrier is 6 MHz above this figure at 805.25 MHz. The oscillator in the tuner would be at a frequency of 39.5 MHz above the vision carrier. This makes

799.25 MHz + 39.5 MHz = 838.75 MHz

Remember:

$$IF = f_{lo} - f_s$$

In this case, f_{lo} is equal to 838.75 MHz, and f_s is equal to 799.25 MHz (vision carrier) and 805.25 MHz (sound carrier).

Let us take the vision carrier first. Additive mixing is used, and we select the difference:

838.75 MHz – 799.25 MHz = 39.5 MHz

This equals the vision IF. The frequency is 39.5 MHz.

Now let us take the sound carrier:

838.75 MHz – 805.25 MHz = 33.5 MHz

This is the sound IF. The frequency is 33.5 MHz.

Therefore there are two IFs generated by the tuner unit (ignoring the chroma IF for the moment), one at 39.5 MHz and the other at 33.5 MHz. These two signals are passed on to the IF amplifier. Note that they are still 6 MHz apart.

ACTIVITIES

1 For a receiver of your choice, connect a voltmeter to the tuning voltage input of the tuner. The voltage range should be within 0–30 V. Now change channels using the channel selector. The voltmeter should indicate different values of voltage for different channels.

2 For this activity you will need a spectrum analyser and a good quality off-air signal. Tune the receiver into the transmission. Connect the analyser into the IF output of the tuner. Tune the analyser to the band of frequencies that corresponds to the IF output: that is, 30–40 MHz. What you should now see is energy (vertical spikes) at approximately 33.5 and 39.5 MHz. These of course are the sound and vision IFs respectively.

The IF stage

The main job of the IF amplifier is to provide: enough **gain** so that the vision demodulator is correctly driven; **attenuation** at certain frequencies, to reduce the effect of interference. Also, owing to the use of vestigial sideband transmission, the vision demodulator will produce twice the output signal for video frequencies below 1.25 MHz. This imbalance must not be allowed to occur.

The required attenuation can be achieved by using special **rejector** and **acceptor** circuits. These 'trap' circuits are usually positioned before the main 'chain' of amplifiers.

Amplification

The vision and sound IF signals receive most of their amplification in the IF amplifier. A typical figure is around 50–70 dB. Transistors connected in the common-emitter mode are used. There may be three or four of them. This arrangement is known as a **cascade** (one after the other).

In more modern receivers, the IF amplifiers may be part of an integrated circuit (IC), or sometimes separate transistors and ICs may be used. The vision demodulator may also be fitted into the same IC.

Figure 14.7 shows the gain/frequency response curve of a typical IF stage. The television receiver is shown to be tuned to channel 62. The IF stage is also known as the **vision IF** stage. Note that the sound IF uses the same amplifiers, and it may also pass through the vision demodulator. It is possible for the sound IF to have its own IF stage, but this creates difficulties if the tuner oscillator drifts in frequency (more on this later).

In Figure 14.7, note that the vision IF (39.5 MHz) 'sits' on the slope of the curve, down by about 6 dB. Hence low video frequencies will be given less gain than high video frequencies. This compensates for the use of vestigial sideband transmission, where the video frequencies close to the vision carrier (0–1.25 MHz) would cause an increase in signal at the demodulator output, as there are two sidebands.

Whenever two signals (the vision IF and the sound IF) share the same chain of amplifiers, it is wise to reduce the amplitude of one of them. This prevents them from interfering with each other. The sound IF uses frequency modulation: so it is sensible to reduce the amplitude of this signal rather than the vision IF. The vision IF uses amplitude modulation. Therefore it sits on the slope at about −30 dB (about one thirtieth of the 0 dB level). This equals about one fifteenth of the vision IF. level.

Two other IF frequencies can also be generated under extreme reception conditions. They are known as the **adjacent channel IFs**. They are unwanted, so it is important that they should be given as much attenuation as possible if they manage to pass through the tuner unit.

In our example in Figure 14.7, the television is tuned to channel 62. The two adjacent channels are channels 61 and 63. Channel 61 has a sound carrier at 797.25 MHz. Channel 63 has a vision carrier at 807.25 MHz. It is these two carrier signals that, if they pass through into the mixer, will beat with the oscillator and produce their own IF signals. This

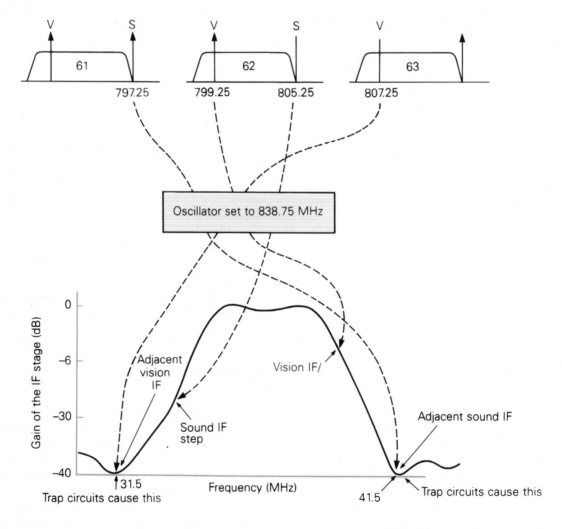

Figure 14.7 Overall IF response of a 625-line (system I) receiver. Receiver tuned to channel 62. Also shown are the two adjacent channels, 61 and 63.

may cause patterning on the screen.

The channel 61 sound carrier will produce an IF of

$$838.75\,\text{MHz} - 797.25\,\text{MHz} = 41.5\,\text{MHz}$$

This is the **adjacent sound IF.**

The channel 63 vision carrier will produce an IF of

$$838.75\,\text{MHz} - 807.25\,\text{MHz} = 31.5\,\text{MHz}$$

This is the **adjacent vision IF.**

As you should now see, the gain at these two frequencies should be at least −40 dB. This is almost one hundredth of the 0 dB position.

Figure 14.8 shows the circuit diagram of a monochrome receiver IF stage complete with the vision detector, which is within IC1. The signal input to the IF stage is via C6 and R8. VT1, VT2 and VT3 are the vision IF amplifier transistors. The collector loads for these transistors are L6, L7 and L9 respectively. **Stray capacitance** exists in each coil, so that each one is effectively a parallel-tuned circuit. These tuned circuits have a high impedance at resonance. Each coil is damped to increase its bandwidths.

L1, C12 and C15 form a trap circuit. This trap attenuates the 33.5 MHz sound IF signal. L3, C16, C18 and C17 is also a trap circuit. It ensures that any signals at 41.5 MHz (the adjacent sound IF) are reduced to such a level that they do not cause any interference.

L5, C22, C20 and C24 also trap any signals at 31.5 MHz. This circuit is called the **adjacent vision IF rejector circuit**. It actually *accepts* the unwanted signals, but this is the same as rejecting or stopping them from being passed through the IF amplifiers. L8, C38 and C35 ensure that the response or gain is kept down at this frequency.

Capacitive coupling is used to pass the signal from one stage to another. C27, C36, and C43 perform this operation.

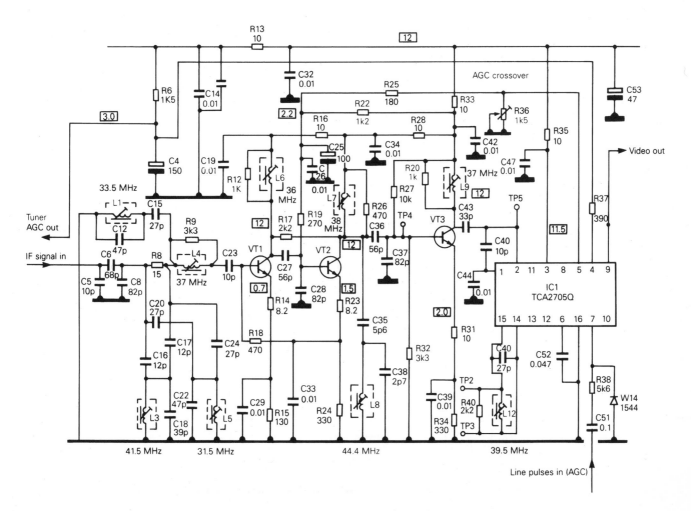

Figure 14.8 IF circuit with the vision demodulator (within IC1).

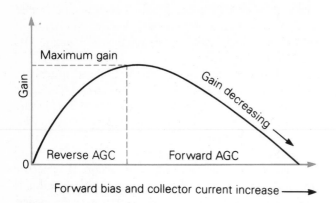

Figure 14.9 Decrease of gain with an increase in forward bias.

Automatic gain control

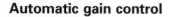

AGC is applied to VT1 and VT2. The VT2 base obtains the AGC voltage from pin 5 of IC1, and controls VT1 in turn from its emitter. **Forward AGC** is used. This means increasing the conduction of the 'controlled' transistors, which will in turn reduce their gain. Figure 14.9 shows the principle of this action.

Staggered tuning

The IF stages need a wide bandwidth and also good gain. Using resonant circuits such as those shown in Figure 14.8

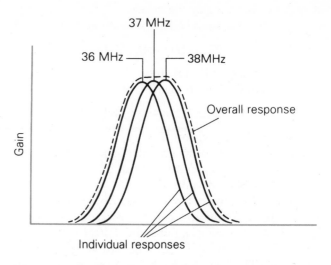

Figure 14.10 The use of staggered tuning for the IF tuned circuits.

Figure 14.11 A surface acoustic wave (SAW) filter, as used to give the desired shape to the IF curve.

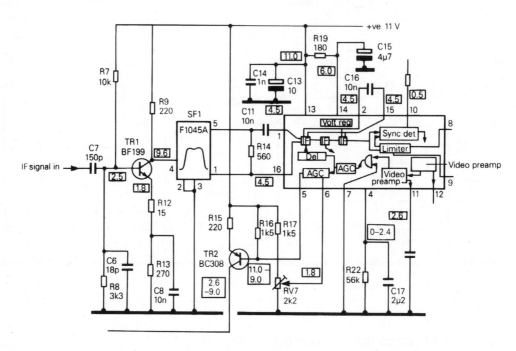

Figure 14.12 Application of the SAW filter, as used in the Ferguson 1600 series chassis.

(L6, L7 and L9) tuned to the same resonant frequency will not give the required gain and bandwidth. Reducing the Q of each tuned circuit increases the circuit bandwidth, but unfortunately the stage gain ends up being small.

A technique known as **staggered tuning** is used to give the required gain and bandwidth. Each stage is given a slightly different resonant frequency to work at: in this case 36 MHz, 38 MHz and 37 MHz. The effect of this is shown in Figure 14.10. By using coils with different amounts of Q and the correct amount of applied damping, different overall shapes may be obtained. The overall response of the complete IF amplifier circuit is shown dashed. To simplify this diagram, the effect of the rejector circuits is not shown.

Surface acoustic wave filters

In modern receivers, conventional tuned circuits have been replaced by surface acoustic wave filter (SAW) filters. SAW filters are reliable, they require no tuning, they are easy to replace, and they are cheap compared with the use of coils and capacitors. Figure 14.11 shows a photograph of one of these devices, which is about 2 cm in diameter. Figure 14.12 shows an IF amplifier circuit with an SAW filter included as the tuning element. It provides all the necessary rejection and response shaping.

The vision demodulator and video amplifier

The vision demodulator's job is to retrieve the original video signal. It also includes a filter circuit, which removes all the IF component of the signal. The vision demodulator is also a convenient place to remove the 6 MHz intercarrier sound signal.

The outputs from the demodulator are:

1. the modulating video signal, bandwidth 0 Hz (d.c.) to 5.5 MHz;
2. the intercarrier sound signal, 6 MHz ± 50 kHz;
3. the chrominance subcarrier, 4.43 MHz ± 1 MHz (for the moment this is included for information only).

The top trace in Figure 14.13 shows the vision IF signal being applied to the vision demodulator (Ferguson 1615 series chassis). The bottom trace shows the demodulated vision IF: that is, the video signal. The chroma and 6 MHz intercarrier sound signals are not shown.

Refer to the circuit diagram in Figure 14.14. For this description we shall refer to the video signal and

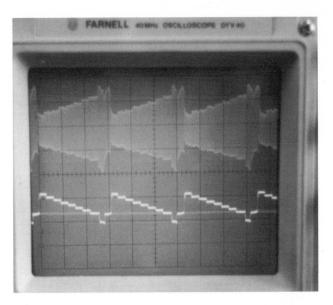

Figure 14.13 Top trace: the IF input to the vision demodulator. Bottom trace: the video signal after demodulation (grey-scale pattern).

intercarrier sound signal only. The sound and vision IFs are fed to the demodulator diode (vision detector) W1 via C26. Coupling is also given by the IF transformer L7b and L7a. This is done to achieve the correct bandpass characteristic (shaping). The top end of L7b, as far as signal is concerned, is at zero potential, owing to the decoupling capacitors C31, C32 and C33. These capacitors are in parallel, not to increase the total value but to ensure good decoupling at all frequencies. (Electrolytic capacitors such as C32 do not work very well at high frequencies.)

W1 and the capacitors C27 and C28 plus L8 look very much like a simple half-wave rectifier. In fact, basically it is. The only difference between a half-wave rectifier mains circuit and this circuit is its operating frequency. C27 is the reservoir capacitor, C28 is the smoothing capacitor, and L8 is the smoothing choke. The load resistor is R23. In this case the values of the filter components are chosen to leave the video signal (0 Hz to 5.5 MHz) and the intercarrier sound signal (6 MHz + 50 kHz) across the load R23.

The diode W1 rectifies the vision signal. This produces a negative d.c. output voltage, which varies with the content of the video signal. To improve the operation of the diode under low IF signal conditions, it is caused to conduct slightly by a small forward bias voltage provided by R24.

As we have already seen, the demodulator's job is mainly to retrieve the original modulation: in this case, the video signal. The demodulator circuit shown in Figure 14.14 is a comparatively simple type of

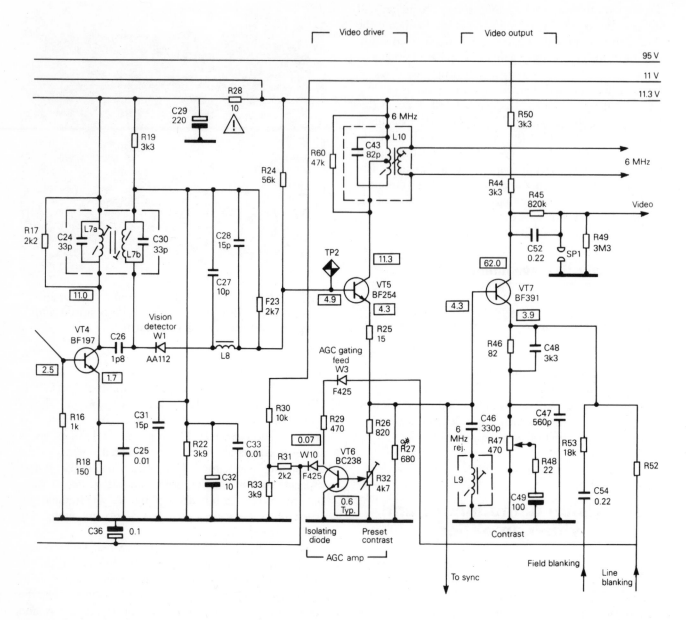

Figure 14.14 Vision demodulator circuit demodulator circuit for a monochrome receiver (Ferguson 1690/91 series).

demodulator. All demodulator circuits, such as the type shown in Figure 14.14, are known as **envelope detectors**. For more information on AM demodulators, see Chapter 5.

The basic type of vision demodulator (envelope detector) uses a diode and a few filter components. The circuit needs to be correctly driven. This means that the IF signal needs to be of a sufficient value for the demodulator to perform adequately. The time constant of the filter components is important so that the charge on the capacitors can follow the required shape of the

video signal. If the time constant is too short, excessive ripple will appear on the video signal. If it is too long, all high video frequencies will be affected. This is because the charge on the capacitors is unable to follow the shape of the video signal (see Figure 14.15).

The vision synchronous demodulator

In a vision synchronous demodulator, the amplitude of the vision IF signal is measured or sampled every half

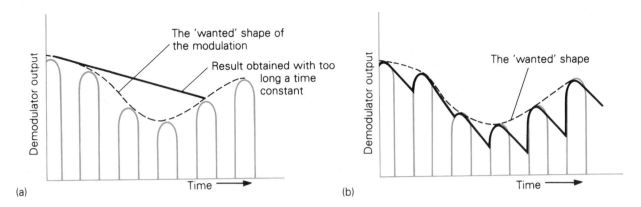

Figure 14.15 The effect on the recovered signal of using incorrect time constants: (a) long time constant; (b) short time constant. The half cycles of IF signal are shown to assist understanding; they would not normally be seen.

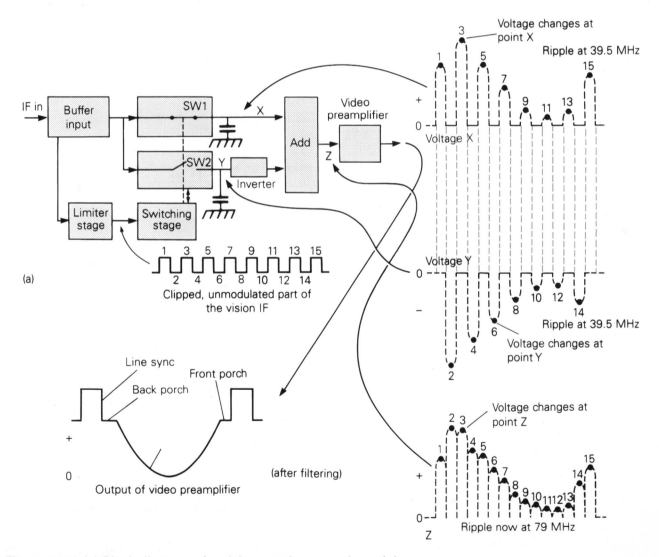

Figure 14.16 (a) Block diagram of a vision synchronous demodulator;

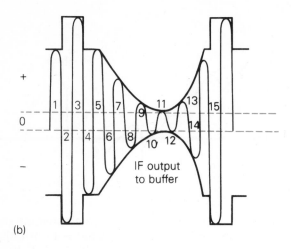

(b)

Figure 14.16 (b) relationship between carrier (IF) and video signal (not to scale).

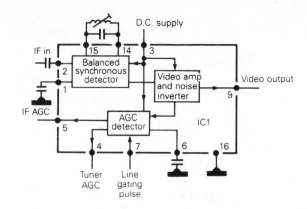

Figure 14.17 Synchronous demodulator integrated circuit, showing the basic functions within the IC.

cycle. The sampling is performed by a switching signal (Figure 14.16). This operates two electronic switches, and closes them at the point when the IF signal is at its maximum amplitude (whether positive or negative half cycles).

The switching signal is derived from the part of the IF signal where there is no modulation: that is, 18–20 per cent above zero carrier (see Figure 10.6). The limiter stage removes all of the video signal and leaves behind a 'clipped' signal, which is used for operating the two electronic switches. The clipped IF carrier is ideal for sampling the vision IF because the frequency and phase of the two are identical

The switching signal is 'switching' its own carrier! When switch SW1 is closed, switch SW2 is open; when SW2 is closed, SW1 is open; and so on. The numbers shown on the switching signal are also shown on the vision IF signal (Figure 14.16a). They show the points on the IF carrier at which each switch is operated. All odd numbers (1, 3, 5, 7, …) mean that SW1 is closed. All even numbers (2, 4, 6, 8, …) mean that SW2 is closed.

Different values of voltages are shown at X and Y. Voltage Y is inverted in the inverter and then added to voltage X in the adder. The resultant video appears at Z. The switching action produces a video signal with a ripple component of twice the IF frequency: that is, 39.5 MHz × 2 = 79 MHz. A 'ripple' of 79 MHz is much easier to filter off as it is much further away in frequency from the highest video frequency of 5.5 MHz. Less video signal corruption takes place.

Figure 14.17 shows a very simple diagram of a synchronous demodulator integrated circuit (IC1 in Figure 14.18). The IF input to the chip is on pin 2. Video output is on pin 9, or pin 10 (not shown). Pin 10 gives the choice of negative-going video if it is required.

AGC processing is also included. AGC to the IF amplifier is applied from pin 5. The AGC to the tuner unit is applied from pin 4. Pin 7 has line pulses on it. These are used for the AGC processing. The tuned circuit shown is to do with the timing of the demodulator circuitry.

Synchronous demodulators do not need a large IF signal input compared with normal envelope detectors, 30–50 mV being usual. They are less prone to demodulating noise on the IF signal. Just a simple resistor capacitor filter circuit is required at the video output. They are very reliable, but cannot usually be made with discrete components as they are too complex. In later types, the same IC may also include an IF amplifier, AFC processing, AGC processing and video of either polarity.

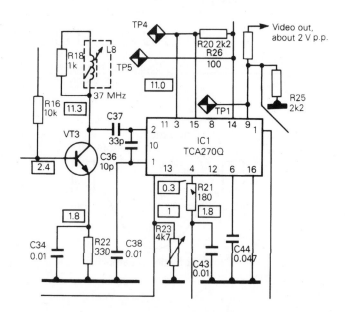

Figure 14.18 Synchronous demodulator IC1 as part of the main circuit diagram (Ferguson 1600 series).

Figure 14.18 shows IC1 as the vision synchronous detector chip. Pins 1 and 2 are the IF input and pin 9 is the video output. Approximately 1–2 V of video signal would be seen at this point.

The video amplifier and buffer stage

The video amplifier used in a monochrome receiver is usually a single transistor operating in the common emitter mode. The bandwidth of this stage has to extend from d.c. or 0 Hz right up to 5.5 MHz. The gain of the video amplifier stage will depend on the size of the CRT, and could be anything between 15 and 40. For a large CRT, say 24 in (61 cm), the change in contrast level (the ratio between black and white) will have to be greater because the changes in brightness are being spread across a larger surface area. Therefore the gain of the video amplifier will need to be greater for a large CRT than for a small CRT.

As already mentioned, the bandwidth has to be reasonably level. Figure 14.19 shows an ideal gain versus bandwidth response curve for a video amplifier stage. The video signals range from d.c. to 5.5 MHz. Any fall-off in gain, especially at high frequencies, will show itself as a lack of definition on the CRT screen: the picture will lack detail. Fall-off in gain at low video frequencies is prevented by coupling the video signal directly to the

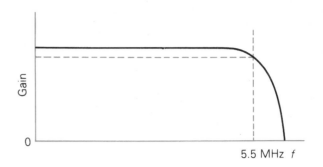

Figure 14.19 Ideal gain versus bandwidth response curve of a video amplifier stage.

CRT cathode: that is, by not using any coupling capacitors. Coupling capacitors would have a high reactance at low frequencies.

To ensure that the frequency response of the video amplifier remains reasonably level, compensation components are added to the circuit. Use is also made of something known as **stray capacitance**. This can exist around the area of the circuit. Receiver manufacturers can make use of this stray capacitance to advantage. Using special high-frequency transistors can also help to maintain a wide bandwidth.

Figure 14.20 shows a simplified circuit of a common emitter transistor amplifier suitable for using as a video amplifier. R1 is the collector load resistor. R2 provides

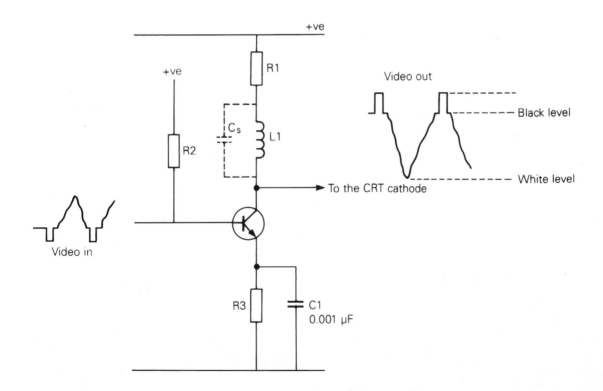

Figure 14.20 A typical video stage.

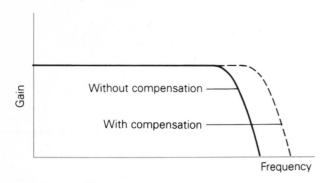

Figure 14.21 Bandwidth of the video amplifier, with and without HF compensation.

the base bias voltage. R3 is the emitter bias resistor and C1 decouples this resistor. L1 is known as a **peaking coil** or **HF compensation coil**.

At low video frequencies L1 does not have any effect, owing to its low reactance. At high video frequencies, the coil becomes (with its stray capacitance) a parallel-tuned circuit. Its resonant frequency will depend on the values of L and C. The effect that this parallel-tuned circuit has on the gain at high frequencies is shown in Figure 14.21. The dotted lines indicate the bandwidth with a certain amount of HF compensation. The full line indicates the response without any compensation.

Many other methods are used, which will give the right amount of compensation to the video amplifier stage. Sometimes a coil is added in series with the video signal. The CRT itself can have inherent stray capacitance. Again, this can be made use of, and this will in the end determine the final amount of compensation.

Negative feedback

If the resistor in the emitter circuit of the video amplifier were not decoupled, it would produce an effect known as **negative feedback**: that is, there would be a reduction in gain. In Figure 14.20 C1 decouples R3. The value of C1 may be anywhere between 820 pF and 4700 pF. At low video frequencies the reactance of C3 would be very high compared with the resistance value of R3. At high video frequencies the reactance of C1 would be very low compared with the resistance of R3. This would give greater decoupling at high video frequencies, thereby gradually increasing the gain.

Many different types of video amplifier circuit are used, some with HF peaking coils and perhaps two or even three different values of emitter decoupling capacitors.

Video amplifiers are voltage amplifiers whose gain varies with variable negative feedback. HF compensation coils are usually used to lift the high-frequency gain.

The video driver

Driving the CRT

The CRT is normally cathode driven. This means that the video signal is fed onto the cathode electrode of the CRT. The video signal has to be inverted: that is, the video part of the signal must be negative going. This will give the correct conditions for the CRT. As the cathode is taken more negative, this will increase the beam current and so give a brighter display for that part of the video signal.

In valve theory, it is normal to show the anode current plotted against the grid voltage (the I_a/V_g curve). The CRT is, in effect, a very big valve: so in this case the anode current becomes the beam current I_b. As the grid is made more negative with respect to the cathode, the beam current reduces, until repulsion of the electrons from the cathode by the negative grid voltage is complete. At this point the CRT is 'cut off'. The beam or anode current is now zero. The same effect can be obtained by making the cathode positive with respect to the grid (the grid is still negative with respect to the cathode).

Television tubes are cathode driven, as shown in Figure 14.22. Cathode drive is more efficient. Grid drive would only give a beam current up to point G for the same amount of signal. The characteristic curve has been drawn as a straight line for simplicity. Some curvature does in fact exist. Figure 14.23 shows the video signal being applied to the CRT cathode.

Driving the cathode with the video signal is known as **cathode modulation**. Driving the grid with the video signal is known as **grid modulation**.

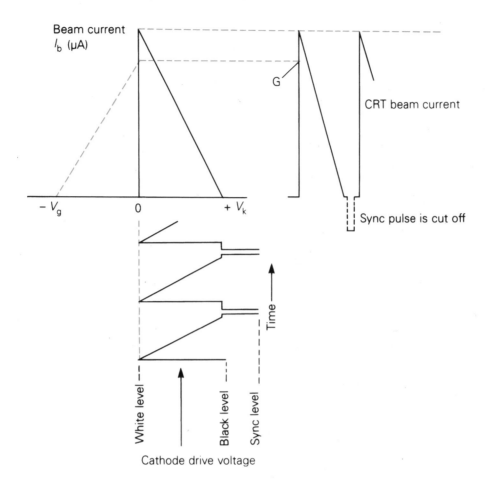

Figure 14.22 The relationship between the beam position on the screen and the transmitted video signal.

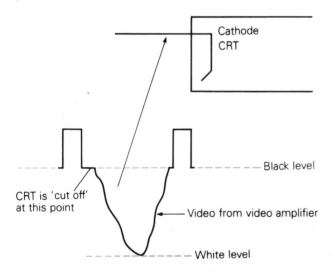

Figure 14.23 One line of video signal being applied to the cathode of the CRT.

Adjustment of the picture contrast

Some adjustment of the picture contrast has to be made available. Changing the picture contrast means changing the ratio between black and white. When we alter the contrast setting on a monochrome or a colour receiver we alter the peak-to-peak value of the video signal being applied to the CRT cathode.

Reducing the contrast **reduces** the ratio between black and white. As a result, the video signal at the CRT cathode will become smaller: the blacks will become grey, and the whites will also turn towards grey. The picture will lack any black or white information.

There are two ways of providing this adjustment. The first is to alter the size of the video signal before applying it to the video amplifier. The second is to alter the gain of the video amplifier stage.

Figure 14.24 shows one way of altering the value of

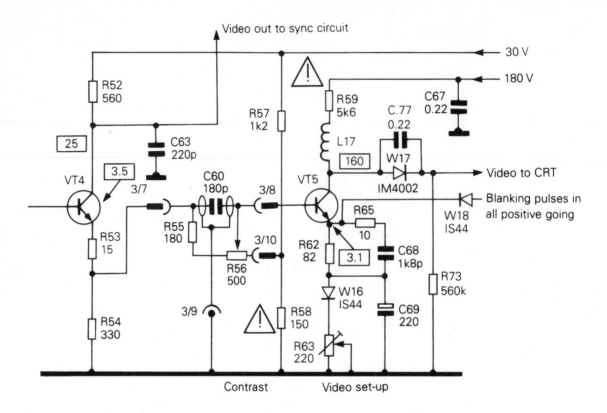

Figure 14.24 VT5 video output stage (Ferguson 1615 series).

video signal at the video amplifier input circuitry. VT5 is the video amplifier. L17 provides HF compensation. The video signal is coupled to the CRT cathode via C77 and W17. When the contrast control is at maximum, the slider of R56 is shifted towards the left-hand side. The video signal travels in from the left of the circuit (from VT4) through R55 and the zero resistance (created by the position of the slider) of R56 to VT5 base. High video frequency signals also pass through the low-reactance C60 and meet up with the main video signals on VT5 base.

If the slider is now moved over to the right-hand side, then the resistance path of R55 and R56 has increased from 180 Ω to 680 Ω. The video signal's path is the same as before, but now there is more resistance put in its way. The video signal being applied to the base of VT5 is now smaller in size, reducing the amount of contrast.

C60 aims to allow high-frequency video signals through, keeping the picture definition high even when the contrast has been reduced.

Figure 14.25 shows another way of controlling the picture contrast in a typical monochrome receiver. Look at the resistor R47, which is in the emitter circuit of the

transistor VT7. This is the video gain, or contrast control. If the slider is at the bottom of the track, C49 and R48 are effectively not in circuit. The gain of VT7 is at a minimum. However, if the slider is at the top of the track, C49 and R48 bypass or decouple R47. (C49 would have a low impedance to signals.) Negative feedback is removed, and the transistor works at full gain. So moving the slider varies the gain and in turn varies the peak-to-peak value of the video signal. Careful choice of values for C47 and C48 also ensures further reduction of negative feedback at higher frequencies to give HF compensation.

SP1 in Figure 14.25 is a **spark gap component**. It protects the video output transistor in the event of CRT flashover. This can occur within the CRT because of the high voltages that are present. Any inter-electrode flashover may cause the cathode connection on the tube suddenly to rise to a very high voltage. This high-voltage transient, if nothing is done about it, may destroy the video output transistor VT7. The spark gap flashes over, thereby preventing the high voltage from presenting itself to the transistor(s).

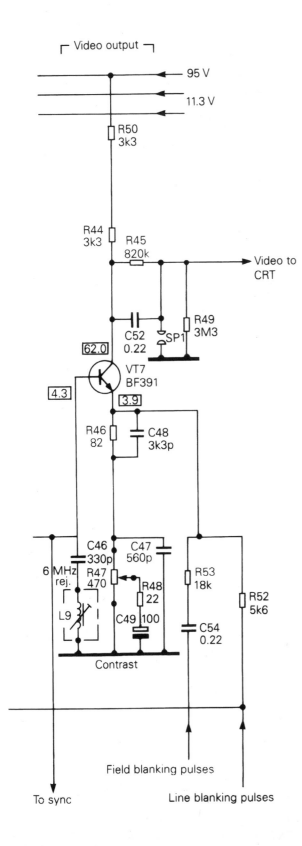

Figure 14.25 Contrast control circuitry and application of the blanking pulses (Ferguson 1690/91 series).

Blanking

The job of the blanking stage is to prevent the beam from being seen during line and field flyback. Some manufacturers employ a separate blanking transistor for this purpose. Others simply feed the blanking pulses directly into the video stage. The blanking pulses can also be fed into the CRT grid circuit. This will give the same result.

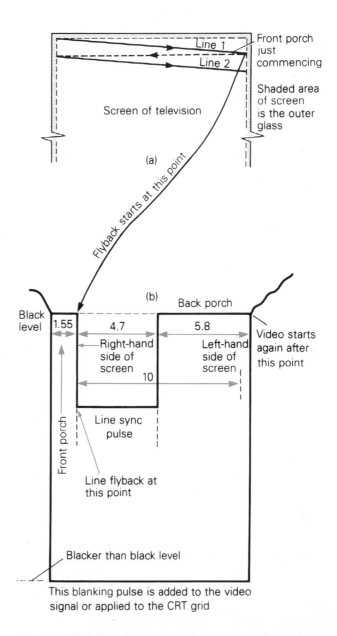

Figure 14.26 Line flyback blanking: (a) blanking pulse shown dotted; (b) relationship between front porch and back porch (all figures in μs).

Line flyback blanking

The blanking of the beam should commence at the end of each line scan: that is, just as the front porch is starting (Figure 14.26). The front porch is at black level. The beam, because of the line timebase, flies back on the leading edge of the line sync pulse. Line flyback takes approximately 10 μs. This brings the beam into the back porch area of the transmitted signal (the last three-quarters of it). A very large pulse can be either added to the video signal, or fed directly to the grid electrode of the CRT. The duration of this pulse is usually about 12 μs.

In Figure 14.25, line-blanking pulses are fed into the emitter of the video output transistor VT7 via R52. These pulses are positive going. Positive-going pulses, if their amplitude is large enough, will turn off VT7, raising its collector voltage. The collector is connected to the CRT cathode. This will turn off the beam in the CRT for about 12 μs. When the beam starts off with a new picture line, the blanking pulse is over. The beam will now be under the control of the video signal. The blanking pulses last for 12 μs but come along every 64 μs. Figure 14.26 shows the relationship between the front and back porch (they are drawn to scale).

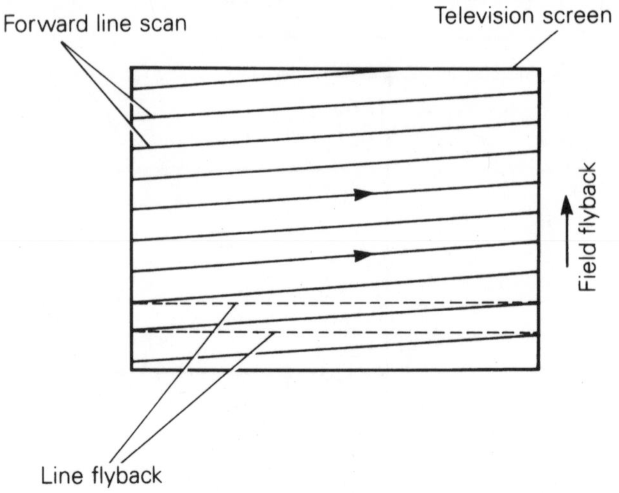

Figure 14.27 Field flyback: what the pattern made by the returning beam could look like. A few line flybacks have also been shown. To see this pattern, the field-blanking pulses need to be disconnected and the brightness control advanced.

Field flyback blanking

Interlaced scanning was described in Chapter 9. It is used in all of the TV systems shown in Appendix 1. In this form of scanning, when the electron beam reaches the bottom of the screen, either on the first (even) field or the second (odd) field, it immediately makes its way back up to the top of the screen. However, because the line timebase (line flyback) is still running, as the beam is moving upwards it is also deflected to the right. This would produce on the screen (if you could see it) a set of sloping lines, similar to those shown in Figure 14.27. The lines are sloping up because the beam is making its way up the screen and not down.

Field flyback takes an average time of 1.8 ms. The pulse needed to blank the beam is approximately the same as this duration. These field-blanking pulses occur every 20 ms.

In Figure 14.25, the field-blanking pulses are fed in via R53 and C54. They are also positive going. They meet up with the line-blanking pulses at the junction of R53 and R52. The line-blanking pulses come from the line timebase and the field-blanking pulses come from the field timebase. If the line-blanking and field-blanking pulses are fed into the CRT grid circuit, they need to be negative going.

Beam limiting

In monochrome receivers, the EHT system is not regulated. It varies about its fixed value with the picture content. For a white screen, the EHT will be at its lowest. For a black screen the EHT will be at a maximum. Some method must be employed whereby the beam current can be limited to a maximum value. If the beam current was allowed to exceed this limit then the picture would suddenly increase in size. This is known as **ballooning**. The picture would also be defocused and lack definition.

To reduce this effect a simple form of **beam limiter circuit** is used. Figure 14.28 shows a basic video output stage connected to the CRT cathode via D4 and C44. VT1 conducts normally, clamping VT1 collector and the CRT cathode to the same voltage. If the video drive signal increases, VT1 collector voltage reduces and also the CRT cathode. The beam current then increases. The current through D4 and R45 decreases. If the signal drive to VT1 is too big the anode voltage of D4 will fall below its cathode and it will turn off. The d.c. component to the CRT will be cut off and now the signal will be a.c. coupled via C44. The black level will now move because of the a.c. coupling and so prevent any further increase in beam current.

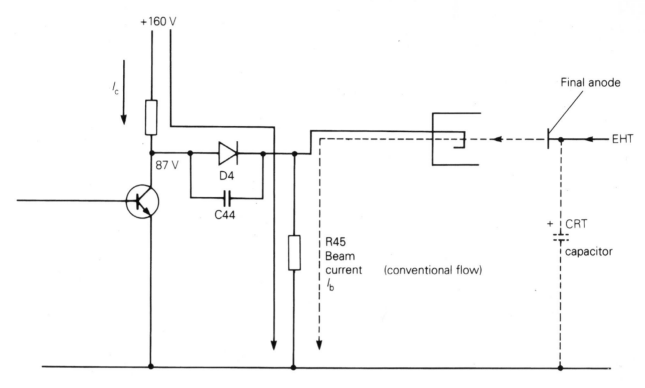

Figure 14.28 The beam limiter circuit.

The video driver stage

The video driver stage fits in between the vision demodulator and the video output stage. The input impedance of the video output stage may be considered to be low in value but changing with different input frequencies. The vision demodulator has a high output impedance. Mismatching will result if the demodulator stage is fed directly into the video output stage. Therefore some form of **buffer circuit** is needed to avoid mismatching. A transistor connected in the common collector mode has a high input impedance and a low output impedance. This transistor mode can be used to match the vision demodulator into the video output stage.

Figure 14.29 shows VT4 to be connected between the vision demodulator (IC1) and the video output transistor VT5. The term 'buffer' is sometimes used for a transistor stage such as VT4, because it stops the demodulator stage from 'looking' into the video amplifier low impedance directly. The emitter load is about $330\,\Omega$ because of the resistor R54. The output from the top of this resistor is directly coupled into VT5 base via the contrast control network. VT4 has no voltage gain. The input and output waveforms are in phase. This is also known as an **emitter follower** stage.

The VT4 stage is also a convenient place to take off an additional signal. A small load resistor (R52) is inserted into the collector lead. From this, the video signal may be developed and then fed to the sync separator.

ACTIVITIES

3 This activity, when set up correctly, will enable you to demonstrate the shape of the IF response curve, as shown in most of the service manuals. You will need a monochrome receiver, a sweep generator or wobbulator capable of covering 30–42 MHz, an oscilloscope capable of being used with the wobbulator, and the manufacturer's service data.

   (The wobbulator is a special kind of signal generator, capable of sweeping over a particular band of frequencies. It is normally used with an oscilloscope. The oscilloscope causes the wobbulator to sweep over the selected band of frequencies in synchronism with the electron beam in the oscilloscope.)

   Connect the oscilloscope to the vision demodulator output. Use d.c. coupling. From this the trace of the oscilloscope will move up (showing an increase in demodulator output) and down (showing a decrease

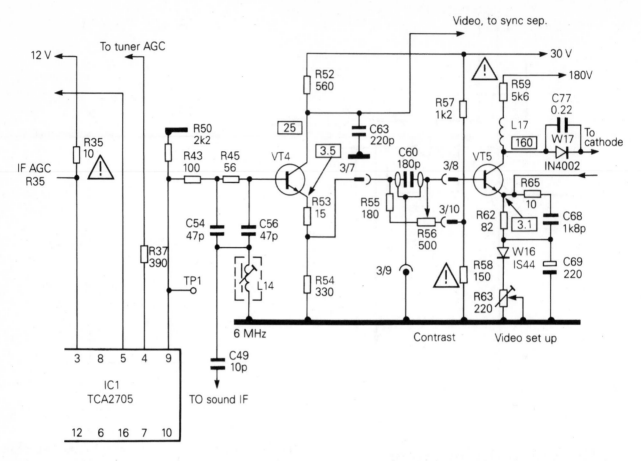

Figure 14.29 The buffer or video driver stage, VT4 (Ferguson 1615 series).

in demodulator output) for different frequencies supplied by the wobbulator. Note that, depending on the type of demodulator (whether the video is positive or negative going), the pattern on the oscilloscope may be upside down. If your oscilloscope has an invert button, all the better! Use the set-up shown in the service manual, or the diagram in Appendix 3.

4 For this activity, you will need a monochrome receiver, the service manual, an oscilloscope, and a colour bar UHF signal.

Locate the video amplifier transistor. Using both channels of the oscilloscope, connect it to the input and the output of the video amplifier. Connect it so that the top trace shows the input and the bottom trace shows the output. Show one line of video signal (timebase set to 10 μs/division). Determine the stage gain, which is equal to

$$\frac{V_{\text{out peak to peak}}}{V_{\text{in peak to peak}}}$$

Depending on the size of the CRT, this figure would be anywhere between 20 and 50. Note that the signal at the output is inverted! If the blanking is performed within this stage, you should see the blanking pulse as part of the video signal at the output.

5 For the same television, locate the blanking stage. Try and find some way of disconnecting it. What we need to see on the screen is the effect of no blanking. When you have achieved this, turn on the TV and tune to a picture. Turn up the brightness control. This will enhance the effect. What do you notice? Is the effect similar to Figure 14.27?

CHECK YOUR UNDERSTANDING

● The job of the UHF tuner is to select, amplify and change the RF signals from the aerial system into more manageable intermediate frequencies.

- The gain of the RF amplifier in the tuner must be automatically variable.
- Varactor or varicap tuning is usually employed. The function of the IF stage is to provide most of the amplification.
- IF shaping is used to prevent interference from the adjacent channels. It also keeps the vision IF and the sound IF at the correct levels.
- For the vision IF, vestigial sideband transmission is used. Therefore the vision IF must be attenuated so that low video frequencies will appear at the correct amplitude relative to the high video frequencies.
- The sound IF (FM) should also be attenuated to prevent interference with the vision IF.
- SAW filters can be used to give the desired shape for the IF channel.
- AGC is also applied; the forward type is universally used.
- The gain must be variable, so that the end user can adjust this ratio.
- Blanking must be included either as part of the video stage or at some later stage at the CRT.
- Blanking prevents the beam from being seen during line and field flyback. It cuts off the beam current for these durations.
- The video amplifier's job is to provide enough gain for the demodulated video signal. It must be large enough to give a good contrast ratio for the picture: that is, from black to white.

REVISION EXERCISES AND QUESTIONS

1 State three jobs carried out by the tuner unit.
2 The IF response curve shown in Figure 14.7 has a complex shape. What is the single component that is now used in modern receivers to give this required shape?
3 UK and many other television transmissions use what is known as negative vision modulation. This means that the vision carrier is negative going: that is, peak white produces a carrier amplitude of about 18 per cent. Why is the carrier not allowed to reduce to 0 per cent?
4 State two advantages of using a vision synchronous demodulator rather than a normal type of envelope detector.
5 An emitter follower or buffer stage normally sits in between the video amplifier and the vision demodulator. Why include this extra stage?
6 A small-screen portable monochrome TV is designed to work from the mains voltage and also from batteries. Most of the circuitry will operate from the low-voltage supply: 12 V. Why is it that the video output stage cannot?
7 What is the approximate amplitude of the video signal at the output of the vision demodulator?
8 The IF response curve is shown in Figure 14.7. Why is the gain of this amplifier low at the two frequencies 31.5 MHz and 41.5 MHz?

The sync separator

Introduction

The sync separator's job is to remove or separate all the sync pulses from the composite video signal. After this operation, the sync pulses must then be separated into line and field sync pulses. The sync separator must ignore the changing video signal. Also, the video signal must not be allowed to filter or leak through the sync separator to the output.

The sync pulses are at black level and continue beyond black level. Figure 15.1 shows the demodulated video signals, complete with line pulses. Also shown is the complicated train of equalising pulses and field sync pulses.

A simple transistor circuit can be used to separate the sync pulses from the video signal. Figure 15.2a shows the circuit of a typical sync separator stage. Capacitor C103 and the base-emitter junction of VT9 form a **d.c. restorer circuit** or **clamping circuit**. The equivalent circuit is shown in Figure 15.2b. Assuming that the diode is ideal (it does not need the 0.6 V across it to conduct), the video signal will sit below the zero line.

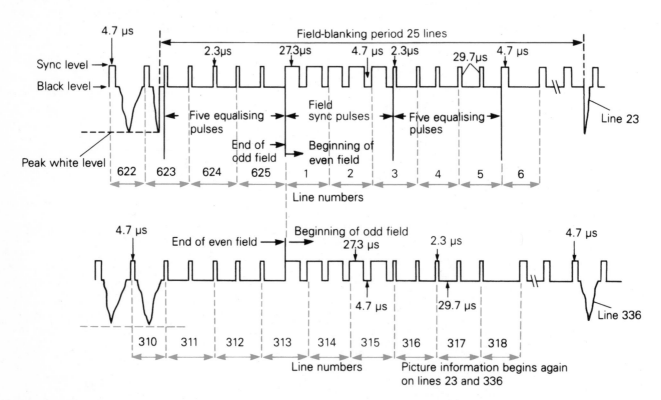

Figure 15.1 Field-synchronising waveforms of the UK 625-line TV system.

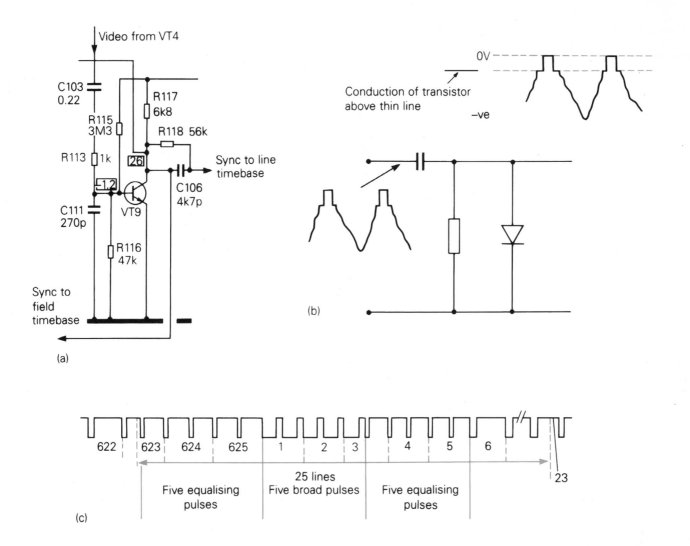

Figure 15.2 (a) A typical sync separator stage (Ferguson 1615 series). (b) Equivalent circuit. (c) Output from the sync separator. All the video signal is removed.

The small positive bias introduced by R115, and the fact that the diode *does* need 0.6 V to conduct, will therefore move the waveform slightly upwards so that the correct 'slicing' point will be reached. The transistor will conduct only above the point indicated. For the rest of the time it will be turned off. The output signal from the transistor will therefore consist of a stream of pulses consisting of line and field sync pulses. These pulses are shown in Figure 15.2c. They are also inverted.

The line and field sync pulses must now be separated so that they can be sent to their respective timebases (the line and field timebases).

Integrators and differentiators

The separated sync pulses (line and field) are now fed to an integrator and differentiator circuit. This circuit provides a simple way of separating these pulses so that they can be diverted to their own timebases. Remember that the sync pulses are needed to synchronise the timebase circuits, so that the electron beam is in the same position on the screen as the beam in the camera (see Chapter 9). The field sync pulse is not one long pulse, but is made up of a number of smaller pulses. This will be explained later.

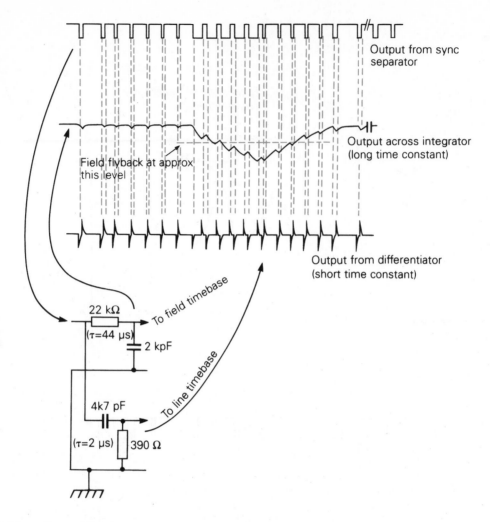

Output from sync separator

Output across integrator (long time constant)

Field flyback at approx this level

Output from differentiator (short time constant)

22 kΩ
(τ=44 μs)
2 kpF
To field timebase

4k7 pF
(τ=2 μs)   390 Ω
To line timebase

Figure 15.3 The effort of the integrator and differentiator on the field sync pulse train.

Integration of the field sync pulse

The train of pulses shown in Figure 15.2c is fed to an **integrator circuit** (Figure 15.3). The time constant of this circuit, τ, is given by

$$\tau = C \times R$$

and should be longer than the duration of one of the field sync pulses (27.3 μs). This circuit is known as the **field sync integrator network**.

As you can see, one large pulse is obtained from the smaller field sync pulses. This happens because of the time constant of Figure 15.3 and the duration of the field sync pulses. Only when the pulse is of sufficient amplitude will the field timebase be triggered. This ensures that noise on the signal will not cause false triggering of the field

timebase. The integrator will not take any notice of short-duration noise pulses.

The same train of pulses is fed to a **differentiator circuit**, which has a short time constant. The output from this circuit is also shown in Figure 15.3. These differentiated pulses are fed to the line timebase.

CHECK YOUR UNDERSTANDING

● The sync or timing pulses must be separated from the composite video signal. The sync separator performs this action.

● The composite syncs must then be diverted to their own timebases.

The field timebase

Introduction

The job of the field timebase is to supply a sawtooth-shaped current for the field deflection coils. This current will create a magnetic field, which will cause the electron beam to be deflected in a vertical direction on the screen: from top to the bottom (scan) and then back to the top (flyback).

The field timebase consists of two parts: a multivibrator circuit, which is attached to a sawtooth voltage generator circuit, and a power stage, which usually consists of a push–pull pair of power transistors. Figure 16.1 shows the block diagram of a typical field timebase.

Certain adjustments must also be included as part of the timebase. These are the multivibrator frequency control (the **vertical hold** control), the **picture height** control, and some form of sawtooth waveform shaping control known as the **field linearity adjustment**.

The field timebase used in a colour receiver has to perform the same job as in a monochrome receiver. The only differences may be that, for the colour receiver, additional waveforms may be necessary for **raster** **correction**, or perhaps (for earlier colour receivers) **convergence** waveforms.

Sawtooth waveform generation

The deflecting coils require a sawtooth-shaped current. The simplest way of doing this is first of all to obtain a sawtooth-shaped voltage. This can be achieved quite easily by charging up a capacitor.

Figure 16.2a shows a resistor and capacitor connected in series. A switch is also connected across the capacitor. When power is applied to the circuit (switch open), capacitor C charges up through resistor R. The time taken for C to charge through R depends on the values of C and R. The time constant = C × R seconds.

Figure 16.2b shows a graph of C charging up through R. The most linear part of the curve lies between the two points x and y. If C was allowed only to charge up to point y, and then the switch was closed, the voltage across C

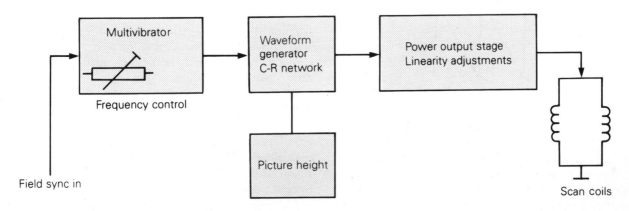

Figure 16.1 Block diagram of a typical field timebase.

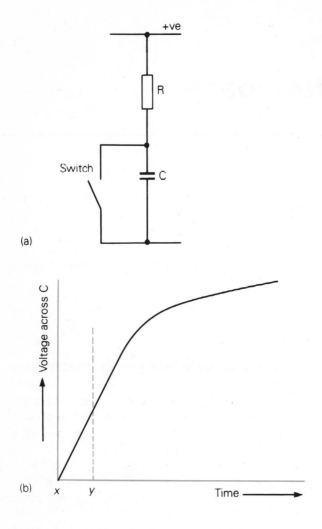

(a)

(b)

(c)

Figure 16.2 (a) A capacitor charging up through a resistor. (b) Graph of C charging up through R. (c) Graph of C charging up through R when the switch is being opened and closed. The switch is closed for a very short time, and opened for a longer time.

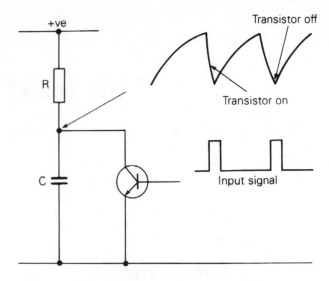

Figure 16.3 A transistor used as a switch.

would fall to zero. Figure 16.2c shows the result of the repetitive opening and closing of the switch.

A transistor can be used as a switch. Figure 16.3 shows this. The transistor can be turned on and off, discharging the capacitor (when on) and letting it charge up (when it is off). The transistor shown in Figure 16.3 can be part of an **astable multivibrator**. The multivibrator has to be free running in the event of there being no field sync pulses. Do not forget that the CR network provides the sawtooth shape, and not the multivibrator itself.

The field oscillator

Figure 16.4 shows the circuit of an astable multivibrator as used in the Ferguson 1615 series of monochrome receiver.

R95 with R96 and C92 and C93 are the components responsible for producing the sawtooth-shaped voltage. R95 is the picture height control. Reducing the value of R95 (slider upwards) reduces the circuit time constant and therefore increases the picture height. These components are called the **field-charging network** or **ramp** charging components.

W23 isolates the field-charging network from the multivibrator circuit. It conducts only when its cathode is less positive than its anode by 0.6 V. This occurs during field flyback. VT7 and VT8 perform the job of the multivibrator. Field sync pulses are fed in via W22. Again, this is an isolation diode, sometimes called an **interlace diode**.

C92 with C93 charges up through R96 and the height control R95. VT8 is turned off; its collector is high. W23 is

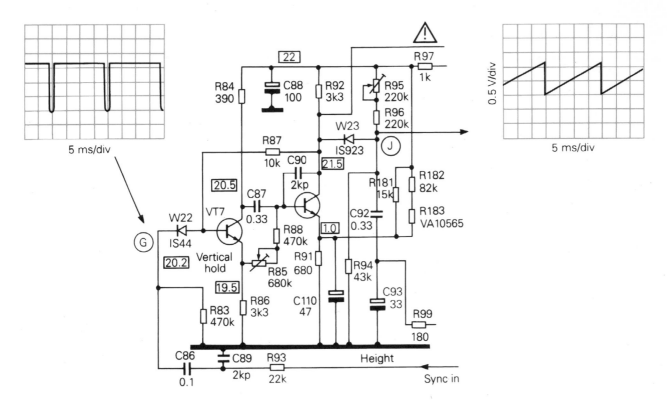

Figure 16.4 The field oscillator, VT7 and VT8 (Ferguson 1615 series).

reverse biased. VT7 is turned on. Base bias is supplied from R87. As C92 is charging up, a linear ramp voltage is being developed between its top plate and the chassis connection (see waveform J).

Integrated field sync pulses due to integrator R93/C89 are applied to VT7 base via W22. This causes VT7 to turn off. VT7's collector goes high. This positive-going voltage on the collector is passed through C87 to VT8 base. VT8 quickly turns on. Its collector drops to a very low voltage and at the same time W23 is turned on.

C92 now discharges through W23 and VT8 and provides field flyback. C87 is charging up through VT8 emitter junction and the resistors R88, R85 and R86. Its right-hand plate is reducing to 0 V. When it reaches about 1.6 V, VT8 turns off. Its collector goes high. W23 turns off. At the same time C92 is now able to charge up again to provide a ramp voltage (scan). When VT8's collector goes high, base bias voltage is supplied via R87 to VT7's base, keeping VT7 on.

When VT7 turns on, its collector comes down from about 20 V to about 3 V. This negative-going voltage is passed through C87 and applied to VT8 base, 'holding off' VT8 for the duration of the scan: that is, C92 charging up. The negative voltage on VT8 base is now allowed to leak away through R88, R85 and R86. When 1.6 V positive is reached, VT8 turns on and allows C92 to discharge via W23 and itself. The process repeats.

R85 controls the discharge time for capacitor C87. It therefore controls the frequency of the multivibrator. It is known as the **vertical hold control**. Waveform J in Figure 16.4 is seen to be a linear ramp voltage at 50 Hz.

Resistor network R182, R181 and thermistor R183 provide thermal stability. They prevent the multivibrator from drifting in frequency, which could occur due to wide temperature changes.

The purpose of W22 is to conduct only when the field

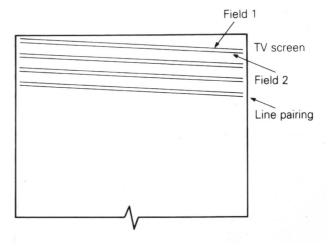

Figure 16.5 The unwanted effects of line pairing.

sync pulse reaches a certain amplitude (see Figure 15.3). This amplitude is known as the **trigger level**. This effect ensures good interlacing: that is, it avoids the effect known as **line pairing**. This occurs when two adjacent scanning lines from different fields 'pair up' (Figure 16.5).

R99 and C93 are part of the feedback network. They are explained in the next section.

The field output stage

The job of the field output stage is to provide sufficient current in the field deflection coils to produce a magnetic field that will cause the electron beam to move down the inner face of the CRT (scan) and also back up again (flyback). The current waveform is essentially of sawtooth shape at 50 Hz.

Quite a large current is needed to perform this action, and so the field output stage is normally a power amplifier. The most commonly used types are known as class B push–pull amplifiers. They usually consist of one or two driver stages, operating in class A, followed by two or four power transistors connected in the common collector mode, and operating in class B.

Figure 16.6 shows a typical field output stage as used in the Ferguson 1615 series of chassis. VT20 is a common-emitter driver stage. VT21 is a current amplifier, which drives the output transistors VT22 and VT23. VT22 is an *npn* type and VT23 is a *pnp* type. They are known as a

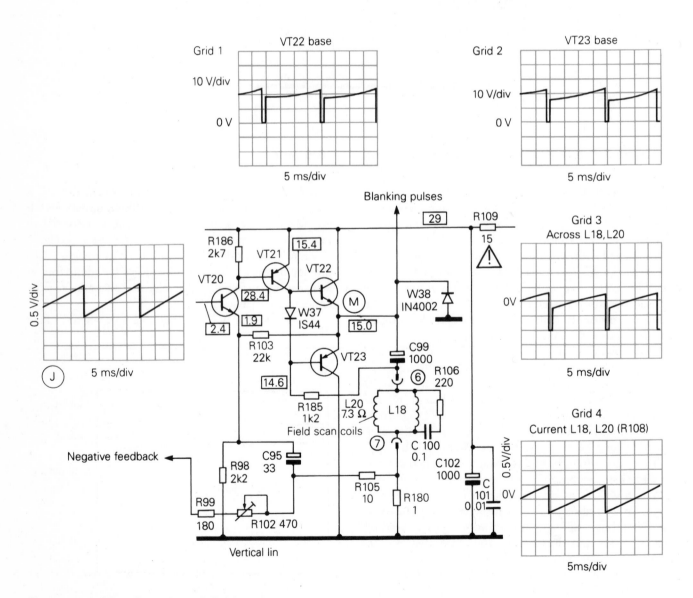

Figure 16.6 Class B push–pull field output stage.

complementary pair, and operate in class B push–pull. The field deflection coils are L18 and L20. They are a.c. coupled to the output transistors via C99. The field oscillator (multivibrator) that supplies this circuit was described in the previous section (Figure 16.4).

VT20 is supplied with a ramp voltage from the field oscillator (waveform J). This signal is then applied to VT21 base. This transistor is essentially a current amplifier, which supplies sufficient current to the output transistors. Notice that the waveform on VT22 base (grid 1) is not of the same shape as waveform J; it is a combination of a rectangular and a sawtooth waveform. The reason for this will be explained a little later in the context of 'S' correction.

Figure 16.7a is an enlarged view of the waveform in grid 1. The dashed line X–X indicates the point at which one transistor (VT22) takes over from the other (VT23). Point B on the waveform indicates the screen centre; point C is the bottom of the screen, and point A is the top. VT22 starts to conduct at point B, pushing current down through the coils via C99. This causes the beam to move from the centre of the screen to the bottom (Figure 16.7b).

When the beam reaches the bottom, the drive waveform on VT23 base (grid 2) (see Figure 16.6) causes VT22 to turn off and VT23 to turn on. This action makes the beam return back to the top of the screen (point A in Figures 16.7a and b). The current in the coils has now reversed and is also at maximum. For this to occur, VT23 current is also at maximum. VT22 is off.

To bring the beam back to the centre of the screen (point B in Figures 16.7a and b), the current in VT23 and the coils must reduce. The drive waveform on VT23 base causes this to happen. VT23's current and also the current in the coils reduces, bringing the beam back to the centre of the screen. Just as the beam reaches the centre point B, VT23 turns off and VT22 turns on. This again takes the beam to the bottom of the screen and the process repeats.

Crossover distortion

VT22 and VT23 operate in class B, or more precisely class AB. This is to overcome a problem that occurs in every circuit of this type: **crossover distortion**. This occurs because of the transistor's inability to function without some forward bias voltage.

Crossover distortion in a field output stage of this type can show itself as a light grey thin band across the middle of the screen. This occurs because the electron beam vertical scan slows down as it reaches the centre of the screen. The scanning lines at this point are compressed together and appear slightly brighter than the normal lines. The fact that the vertical motion of the beam has

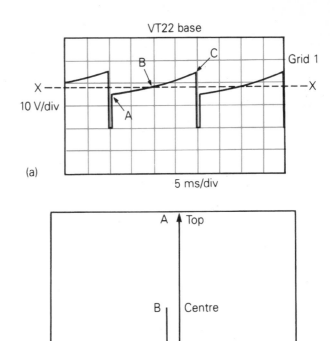

Figure 16.7 Transistor drive waveform for the output transistors; (b) movement of the beam by the field timebase, assuming that the line timebase is not operating.

effectively 'stopped' in the middle reinforces this effect. Examination of the drive waveform across the deflection coils would reveal a small kink appearing at the centre point.

VT22 and VT23 must be biased slightly 'on' to overcome crossover distortion. There should be a **quiescent** current flowing. This explains why the two base connections for these transistors are not simply connected together. The diode W37 causes slightly different base bias voltages to occur for each of the transistors.

Drive waveform

The field deflection coils are mostly resistive, but with a small amount of inductance. At 50 Hz, the coil's reactance is small, because $X_L = 2\pi f L$. The effective circuit of the deflection or scanning coils is shown in Figure 16.8a. The resistance of the coil is shown as a separate component,

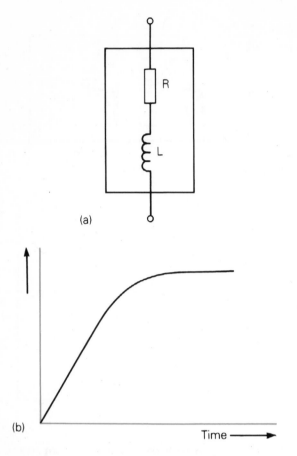

Figure 16.8 (a) The effective circuit of the scanning coils; (b) graph of *I* increasing through L.

and therefore the coil itself can be considered as a pure inductor.

A resistor is non-reactive (as long as it is not a wirewound type), unlike a coil or a capacitor. It is quite easy to obtain a sawtooth-shaped current in a resistor. It can be done by feeding a sawtooth-shaped voltage across the resistor. Obtaining a sawtooth-shaped current in a coil is more of a problem. The coil will oppose any increase in current due to the back e.m.f. Figure 16.8b shows the growth of current in a coil; it can be seen that the current rises slowly. The time constant τ of the coil can be determined by the formula

$$\tau = \frac{L}{R}$$

where L is the coil's inductance and R is its resistance. The graph in Figure 16.8b is similar in shape to the voltage build-up on a capacitor in an integrator circuit.

The deflection coils themselves cannot be used as a current integrator because their inductance is too low. A full time constant (4×20 ms) of at least 80–100 ms would be required to produce a linear scan for about 20 ms.

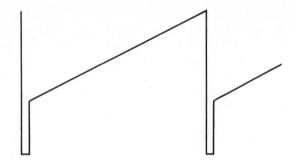

Figure 16.9 Ideal shape of the drive waveform: a combination of a sawtooth and a rectangular waveform.

Therefore the scan drive waveform is modified from a simple sawtooth to a combination of a sawtooth and a rectangular waveform. Figure 16.9 shows the correct drive voltage waveform that is supplied to the field deflection coils.

Grid 4 in Figure 16.6 shows the shape of the resultant voltage across the 1 Ω resistor R108. This results from the current flow in R108 and the deflection coils. As the resistor R108 is non-reactive, this is also the shape of the current waveform in the coils. Notice that it is basically a sawtooth-shaped waveform.

S correction

Figure 16.10 shows the theoretical shape of the current waveform in the field scan coils. It is perfectly linear from A to C. In practice this shape cannot be used; it has to be modified because the screen is almost flat. The deflection centre (the point in the tube from which the beam is deflected) does not correspond to the centre of curvature. A nearly flat screen would in fact require a

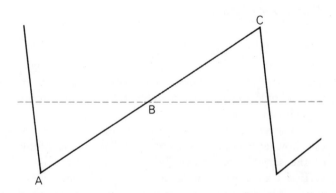

Figure 16.10 Theoretical shape of the current in the scan coils.

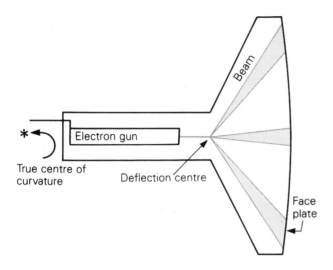

Figure 16.11 Side view of a CRT, demonstrating field scan.

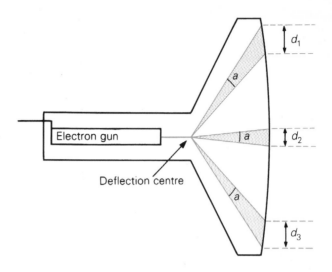

Figure 16.12 Side view of a CRT. d_1, d_2 and d_3 are the distances on the screen caused by the deflection angle *a*. The increase in distance of d_1 and d_3 is caused by the same deflection angle *a*.

very long radius of curvature. Figure 16.11 shows a side view of a CRT, and indicates the deflection centre. The true centre of curvature of the screen would be somewhere to the left of this diagram. Making a CRT of this length would not be practical, and it would create difficulties in its operation.

Figure 16.12 again shows a side view of a CRT. The electron beam is shown in six different positions. Scanning is from top to bottom (ignoring line scan for the moment). The angles *a* are all equal. This downward movement of the beam is created by the linear sawtooth current, but the distances covered on the screen face are not the same. Although d_1 and d_3 are equal, d_2 is smaller. The deflection angles are the same, but this has produced different distances on the screen face. It would appear that the electron beam has moved faster for distances d_1 and d_3 and slower for distance d_2. Geometric shapes such as circles in the scene would appear 'egg shaped'. The top and bottom would appear stretched.

What needs to be done is to alter the scanning current's rate of change. We need to distort the shape of the scanning current so that unequal scanning angles correspond to the same distance covered on the screen face. Figure 16.13 shows the linear scanning current in dotted lines, and the modified shape of the scanning current in solid lines. The rate of change for A and C is now less than for the linear shape, while the rate of change for B is greater. The effect that this waveform will have on the beam will be to slow it down at the edges (d_1 and d_3) and speed it up in the middle (d_2). The result of this modification should be to produce a linear picture on the screen.

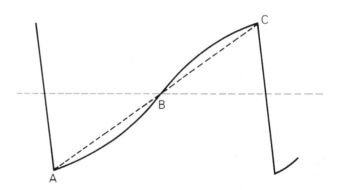

Figure 16.13 S correction. The dotted line shows the linear-shaped current; the solid line shows the S-corrected current.

This type of correction is called S correction, because the shape of the scanning waveform resembles a slanted letter S (see grid 4 in Figure 16.6)

Field linearity controls

Other types of distortion can alter the desired shape of the current waveform. They include such things as changes in the bias conditions for the power transistors, and the scan coil inductance. The scan currents may be 'transformer coupled' to the deflection coils: this can introduce distortion, as can the replacement of certain components, such as transistors and capacitors. Some adjustment must be provided so that such factors can be allowed for.

Field linearity controls are provided to tailor the final shape of the scan current. They are usually connected as part of a negative feedback circuit. Resistors and capacitors can be used in this feedback path to alter the final result. Parabolic waveforms are produced with integrator networks. Adding these waveforms to the linear ramp voltage will change the overall shape.

Figure 16.6 shows the negative feedback path from the top end of R108 through R105 and through the vertical linearity control R102. This path leads back to the field charging network (C92/C93) in the oscillator circuit (Figure 16.4).

The integrator R99C93 provides waveform shaping to the 'feedback' ramp waveform. Integrating a sawtooth waveform at the junction of C92 and C93 gives a parabolic waveform. This is the ideal shape to allow the voltage at the top plate of C92 to build up in a special non-linear way – that is, giving an S-corrected shape.

Later receivers

In later types of receiver, the field timebase is usually included in a single IC. The only components exterior to this may be the scan coil's coupling components, field charging network etc.

ACTIVITIES

1 For this activity you will need an oscilloscope, a monochrome receiver, service data, and an off-air transmission.

Connect the oscilloscope to the top end of the deflection coils. Set the timebase speed to about 5 ms/division. The trace should now be displaying the voltage across the deflection coils. It should be similar to Figure 16.6, grid 3. A slight amount of S correction should be noticeable with this waveform. Now, monitor this waveform while making adjustments to (a) the **height control**; the **vertical linearity control**. What happens to the waveform in each case?

2 This activity will show you the correct way to adjust the field timebase to set the field geometry. In its entirety, it is suitable only for monochrome receivers. In addition to the equipment in Activity 1, you will

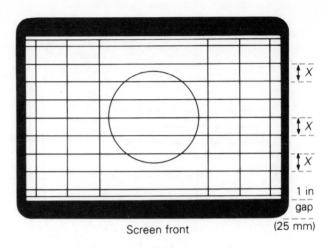

Figure 16.14 Setting the field geometry. The picture should be adjusted so that all the *x*s are equal.

need a pattern generator capable of supplying a pattern similar to the one shown in Figure 16.14, or a pattern similar to test card F (Appendix 2).

Reduce the picture height, using the relevant control, so that about 1 in (25 mm) of blackness is visible at the top and bottom of the screen. Using the field linearity control(s) (there may be two of them), linearise the picture so that the distance between all of the horizontal lines on the pattern is the same throughout the field scan (see Figure 16.14). If the picture does not appear to be in the correct place on the screen, you can centre it on the screen by using the picture shift controls (the ring magnets). When you have centred the picture, increase its height to the correct level. The picture should now be reasonably linear, of the correct height, and also in the correct place on the screen.

For a colour receiver, you can use all the above method apart from shifting the picture around the screen. In a colour receiver the ring magnets perform a different function. In older tubes they were used to adjust the colour purity; in later tubes they altered the purity and the beam convergence. For these tubes, picture shift was achieved electronically by passing a steady d.c. current through the line and field scan coils. The scan drive currents (sawtooth) were added to these d.c. currents. The effect of this steady current was to alter the position of the deflection centre for the electron beams. The value of the current could be varied to alter the position of the picture on the screen. These controls are known as **line** and **field shift** controls.

CHECK YOUR UNDERSTANDING

● The field timebase must provide sawtooth currents in the deflecting coils to give vertical scan.
● The current in the coils must also be S corrected to cope with scanning a nearly flat screen.

1 In any television receiver, why is the sawtooth current used for scanning not perfectly linear?
2 Explain the meaning of the term 'crossover distortion'.
3 How is S correction achieved in the field timebase of a monochrome or colour television?
4 Why is S correction necessary?

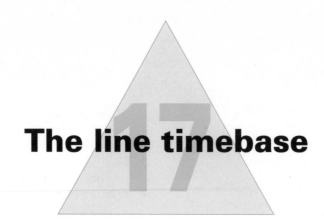

The line timebase

Introduction

The function of the line timebase is to cause the electron beam to scan from left to right across the inner face of the CRT. The frequency of the timebase for the 625-line system is 15 625 Hz. About 9–11 μs is allowed for line flyback. This allows between 55 and 53 μs for scan. Line scan and flyback together always equal 64 μs.

As with the field timebase, a sawtooth-shaped current must be available for the line scan coils. S correction must also be provided. The line timebase can also be used to supply the extra high tension (EHT) that is needed for the CRT. Additional supply lines can also be supplied by this timebase circuit.

Timebase circuitry

Figure 17.1 shows the block diagram of one type of line timebase used for a monochrome receiver.

As for the field timebase (Chapter 16), the line oscillator cannot be directly synchronised: the line sync pulses cannot be fed directly to the line oscillator because of the effect of short-duration interference pulses. The 625-line system uses negative vision modulation. Therefore the effect of random interference could be to 'wipe out' a number of transmitted line sync pulses. This sort of interference does not usually affect the field oscillator because of the relatively long time constant of the field sync integrator network. Interference pulses are usually of short duration, and this circuit usually ignores them.

The line sync pulses originate from the sync separator via some form of differentiator circuit (see Figure 15.3). This usually has a short time constant. Therefore any form of short-duration interference could pass straight through and upset the line oscillator.

The line oscillator is normally under the control of a circuit that has a 'flywheel' action: that is, it has **momentum**. Like a mechanical flywheel attached to a rotating shaft, it will tend to run at the correct speed. If a small number of line sync pulses were missing or obliterated (say 1 in every 12) then the oscillator would not suffer, and would still run at the correct frequency. This technique is called **flywheel synchronisation**.

The line oscillator must run at the correct frequency,

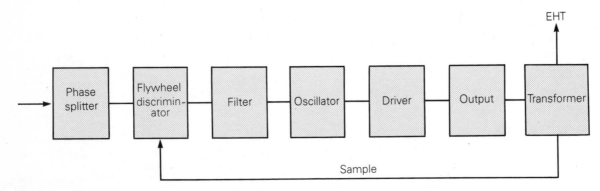

Figure 17.1 Block diagram of a line timebase.

otherwise the picture will be displaced horizontally. Also, in a colour receiver, gating circuits are operated from line pulses derived from the line output stage. Any deviation in the oscillator frequency will produce errors in the timing of these gating circuits.

Figure 17.2 is a breakdown of the line timebase, showing only the front end of the main block diagram.

Figure 17.3 is part of the circuit of the Ferguson 1600 series. VT12 is a phase splitter. Line sync pulses entering at the base will appear at the emitter and collector equal but opposite in sense: negative going at the emitter and

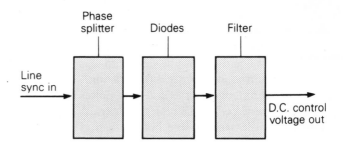

Figure 17.2 Simplified front end of the line output stage.

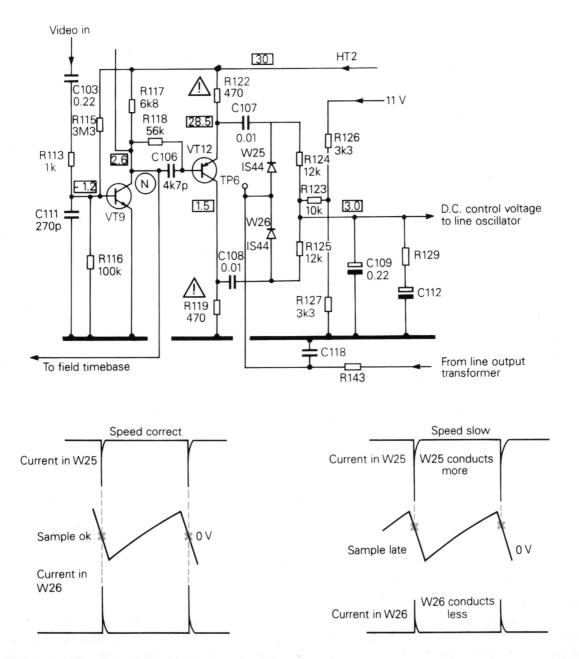

Figure 17.3 Part of the line timebase (Ferguson 1600 series).

positive going at the collector. The line sync pulses are differentiated by C107 and C108. Diodes W25 and W26 are connected across the right-hand plates of the two capacitors. When these two diodes conduct, a short time constant is introduced: hence differentiation.

The leading edge of the line sync pulses causes the diodes to conduct. The circuit is symmetrical: that is, the top half is the same as the bottom half. The current through each diode and capacitor is equal. The voltages across R124 and R125 are equal but opposite in polarity (ignoring R123). In effect, the voltage at the junction of R124 and R125 is at zero potential with respect to the chassis. Under these conditions, this circuit produces no output.

Fed in at the junction of W25 and W26 (TP6) is an integrated flyback pulse. This waveform is shown in Figure 17.3. As you can see, it resembles a sawtooth shape. The integrator circuit is R143 and C118. The frequency of this sample waveform is of course 15 625 Hz, and it is representative of the frequency and phase of the line oscillator circuit. Any drift in the oscillator frequency will also alter the frequency and phase of this signal.

The timing of this waveform should coincide with the current flow through the two diodes, when the flyback portion of this waveform passes through zero. If the frequency of the oscillator is correct then this circuit will produce zero output. Figure 17.3 shows this to be so. If flyback occurs too late, then the sampling point will also occur late. Circuit imbalance will result: W25 will conduct more than W26. The circuit will produce a positive error voltage at the junction of R124 and R125. This condition will mean that the oscillator is running too slow.

Any deviation in the speed of the line oscillator will produce an error voltage at the junction of R124 and R125. The error signal at the junction of R124 and R125 will swing above and below zero. The error signal produced is used to control the frequency of the line oscillator. Any variation in the oscillator frequency will be corrected by this error signal voltage.

It is not practical to control the frequency of an oscillator with an error signal that can swing above or below 0 V. Therefore a small positive voltage is fed in via R123 from the potential divider R126 and R127.

The voltage at the junction of R124 and R125 is about 3 V. Any error voltage produced from the diode circuit will add to or subtract from this 'standing bias': that is, it will swing above or below depending on the correction needed. The voltage at the junction of R123 , R126 and R127 can be made variable to act as a set frequency control. Any correction voltage will of course swing above or below this preset value.

This diode circuit is known as a **flywheel discriminator** or **coincidence detector**.

The d.c. voltage output from the detector is smoothed by C109. This solitary capacitor provides the flywheel action. A damping filter is also used. R129 and C112 provide for this, and prevents **hunting**. This is the term used in any sort of feedback or control circuit where instability may occur. Insufficient damping in this area will cause the oscillator to increase and decrease in speed until it finally settles down. This would be noticeable on the screen as displaced scanning lines. The effect would be worse at the top of the screen than further down.

Many different types of flywheel circuit are available. Transformers were once used to perform the phase-splitting action. Sometimes the line sync pulses are fed directly to the flywheel diodes. The end result is still the same, in that all such circuits provide a flywheel action, and a correction voltage is produced to control the oscillator speed. In later receivers, the discriminator together with the line oscillator is usually part of an integrated circuit.

The complete flywheel-controlled line timebase is a major improvement on the early circuits of the 1960s and 1970s. However, good as it is, it can have a disadvantage when the receiver is used in conjunction with a videotape recorder. Head-to-tape speed is not perfect in any mechanical system. Variations will exist, and this could cause the line sync pulses to occur at different intervals. If the oscillator is under the control of a stable flywheel sync stage, then it might not be able to react quickly to these timing errors. The solution is to modify the reaction time of the flywheel circuit so that it *can* respond quickly. The general modification is to reduce the value of the damping filter capacitor to about two-thirds of its normal value. This will increase the response time of the circuit. Modern receivers have a dedicated video channel. When selected, this channel automatically switches in a capacitor, which will reduce the overall value.

The line oscillator

Next, we shall look at the line oscillator circuit, and how the d.c. control voltage from the flywheel circuit actually controls the frequency of this oscillator.

Oscillator circuits such as multivibrator or even blocking oscillators have been used in the past, but receivers designed for the past 20 years have used a type known as a **Hartley oscillator**. The tuning element in this oscillator is an LC circuit. An oscillator that utilises an LC circuit is usually quite stable in its operation, and suffers less from frequency drift.

This type of circuit will normally generate a waveform of sine wave shape. As we shall see later, the line output stage requires a drive waveform that is basically rectangular in shape and not sinusoidal. A Hartley oscillator circuit can be made to generate a rectangular shape rather than a sine wave shape.

Remember that the d.c. control voltage from the flywheel discriminator has to be able to control the frequency of the oscillator circuit, and that the oscillator circuit has for its tuning elements an L and C circuit. Therefore the d.c. control voltage has to effectively alter one of these properties.

In the tuner unit, varicap diodes are used to control the tuning of the oscillator and of the tuning of the RF amplifier. However, they cannot be used to alter the frequency of the line oscillator, because the amount of capacitance change would be insufficient to have any overall effect. The frequency of the line oscillator is quite low (15 625 Hz) compared with the frequency of the oscillator and RF circuits in the tuner (hundreds of MHz). Therefore some other way must be found to achieve this.

Reactance stage

An additional circuit is introduced so that it forms part of the line oscillator circuit. It is known as a **reactance stage**, and it can act as a variable capacitor. The amount of capacitance depends on the d.c. current passing through it. A transistor is used for this purpose; the amount of d.c. current alters its conduction and hence the reactance.

Figure 17.4 shows an example of a reactance and oscillator stage. VT13 is the reactance transistor. VT14 is the oscillator transistor. The tuning elements for the oscillator are the right-hand side of L19 and C114. The centre of L19 is at earth potential as far as signal is concerned. Base and start-up bias is provided by R134. VT13 and associated components are connected to the left-hand side of L19. Positive feedback for oscillation is fed back from the right-hand side of L19 to the left-hand side of L19 (the two coils are inductively coupled). The transistor VT14 is driven between saturation and cut-off, developing an almost square-wave signal at its emitter. So even though the Hartley oscillator is basically a sine wave type of oscillator, we do not need a sine wave output but rather a more square waveform. This condition suits the line output transistor.

C115 and R133 provide almost a 90° phase shift. Therefore the voltage across R133 represents the current through the capacitor: that is, the current leads the voltage across C115.

The voltage across the resistor is injected into the emitter of VT13 via C116. Figure 17.5a shows the voltage at the collector of VT13; Figure 17.5b shows the voltage at the emitter. The waveform in figure 11.54b must also be the current waveform within the transistor. As can be seen, there is a phase shift of approximately 90° between the two waveforms. The emitter voltage/transistor current waveform is leading the collector voltage. If there was a capacitor across the left half of L19 the same voltage/

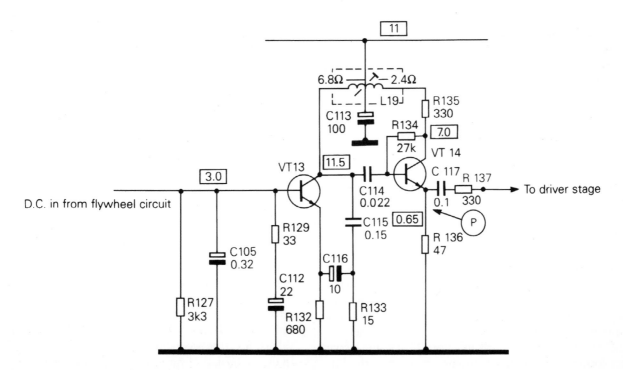

Figure 17.4 The line oscillator in the Ferguson 1600 series.

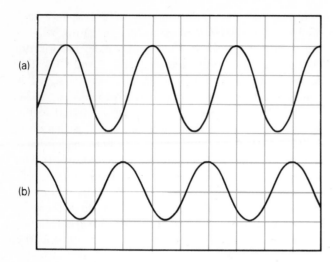

Figure 17.5 The voltages at (a) the collector and (b) the emitter of VT13.

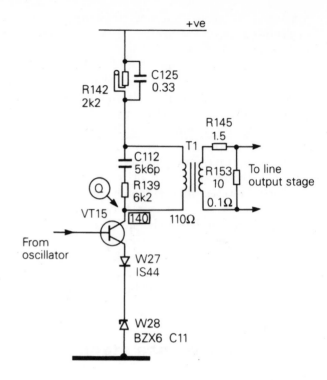

Figure 17.6 The line driver stage of the Ferguson 1600 series chassis.

current relationship would apply, but VT13 is across that coil. Therefore VT13 'looks like' a capacitor. Altering the current through the transistor by altering its base voltage also alters the phase shift between the two waveforms. This effect simulates a change in capacitance, and as this capacitance is effectively across the main tuning element, the frequency of the oscillator can be varied.

L19 is variable. It can be used to alter the frequency of the oscillator, and must be adjusted in accordance with the manufacturer's instructions.

The line driver stage

The line driver stage basically amplifies the output signal from the line oscillator. It may also provide shaping of the drive waveform. Impedance matching is also a function of this stage.

Figure 17.6 shows the circuit of a typical line driver stage (Ferguson 1600 series). VT15 is the line driver transistor. Transformer coupling is used. The transistor is turned on and off by the drive waveform from the oscillator. When it is turned off, the back e.m.f. created by the inductance within T1 may cause damage to the transistor. Therefore a damping circuit is required to prevent this happening. C122 and R139 perform this function. T1 is a step-down transformer, providing a high current drive to the line output transistor.

The line output stage

The main job of the line output stage is to cause the electron beam to be deflected across the screen (scan) and also back to the start (flyback). It does this by generating a sawtooth-shaped current in the line-deflecting coils (line scan coils). The method used to achieve this is different from the method used in the field timebase. In the line output stage the scan coils have a long time constant. Therefore the coils can be used to develop a linearly rising current: that is, they can act as an integrator circuit. S correction must also be applied to the line scan current.

The line output stage can also be used to supply extra facilities, such as the CRT heater current and high tension supply lines.

The simplified circuit of a line output stage is shown in Figure 17.7a. Assuming that there is no field scan, the basic rules of operation are as follows. When the beam is in the middle of the screen, the current in the deflecting coils is zero (Figure 17.7b). When the beam has moved to the far right-hand side, the current has increased to its maximum value. This is the second half of the line scan (Figure 17.7c). In Figure 17.7d, the beam has moved back to the middle of the screen. The current in the coils is zero. This is the first

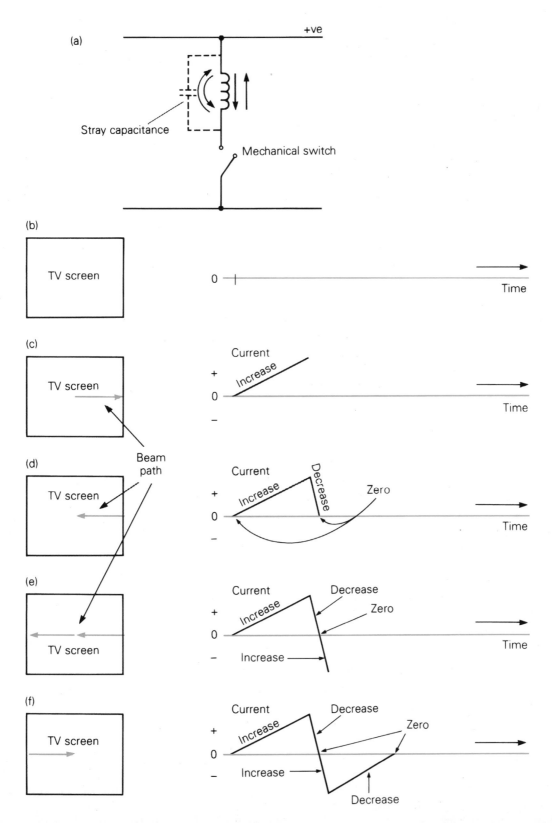

Figure 17.7 (a) Simplified circuit of a line output stage. Current in scan coils: (b) beam in middle of screen; (c) second half of scan; (d) first half of flyback; (e) second half of flyback; (f) first half of scan. After this, the process repeats from (c).

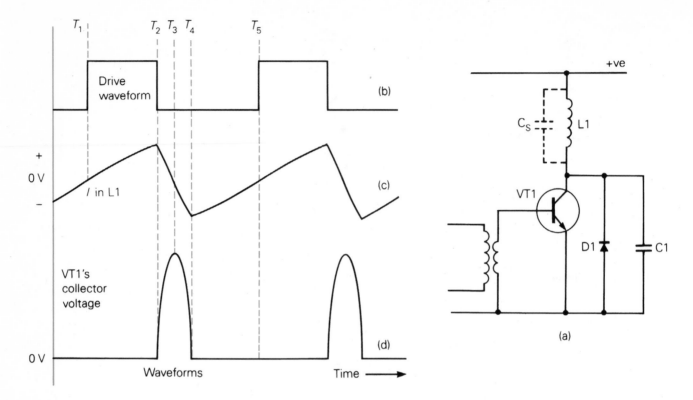

Figure 17.8 (a) A basic line output stage. Waveforms: (b) drive waveform; (c) current in L1;   (d) VT1 collector voltage.

half of flyback. In Figure 17.7e, the beam has now moved from the middle to the far left of the screen. The current has reversed direction and has also increased to maximum. This is the second half of flyback. In Figure 17.7f, the beam has moved from the left-hand side back to the middle of the screen. The current now reduces from its maximum reversed value back to zero. This is the first half of the scan: that is, the beam has travelled from the left to the middle.

We shall now look at the operation of a basic line output stage (Figure 17.8a). Bear in mind the simple rules listed above. For this circuit a transistor takes the place of the mechanical switch.

The drive waveform shown in Figure 17.8b is applied to the base of VT1. At time T_1–T_2, the transistor conducts heavily: it 'bottoms'. All the supply voltage is applied across the coil L1. The current increases to maximum (Figure 17.8c), and the beam moves from the middle of the screen to the right-hand side.

At time T_2 transistor VT1 turns off. No instantaneous voltage appears across the coils. The magnetic field around the inductor collapses. This collapsing magnetic field induces a voltage in the coil (back e.m.f.):

$$e = -L\frac{di}{dt}$$

At T_3, the current in the coil is now at zero. The beam is in the middle of the screen; the first half of flyback has taken place.

During the time interval T_3–T_4, C1 and C_s now discharge through coil L1. The current changes direction, and quickly increases to its maximum value, at T_4. The beam is on the left-hand side of screen. At T_4, the capacitors will be fully discharged. The current will be at a maximum. The magnetic field now collapses linearly and induces a constant voltage in the coil (see formula above). The direction of this induced e.m.f. is opposite to that produced in the time interval T_2–T_4. The collector voltage therefore tries to swing negative. D1 conducts to prevent this happening.

During the time interval T_4–T_5, the current in L1 and D1 is allowed to reduce linearly to zero. At T_5, VT1 is switched on again by the drive waveform in Figure 17.8b. The process repeats. VT1 provides the second half of scan. The recovered energy from the coils is used to provide the first half of the scan.

C1 is known as the **flyback tuning capacitor**. Its value is critical, in that with the stray capacitance (C_s) and the inductance of L1 the resonant frequency of the circuit can be decided. If there is too little capacitance at this point, the resonant frequency of the circuit will be

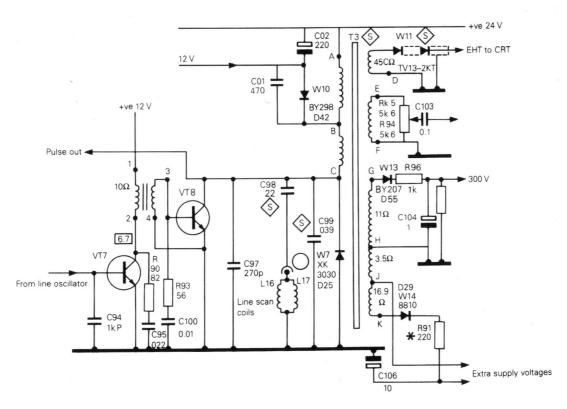

Figure 17.9 Circuit of a typical line output stage, as used in a small-screen portable receiver (Ferguson 1600 series).

increased. This could cause the voltage at the collector to rise beyond the maximum working voltage of the transistor. C1 is usually an unusual value of capacitor with a high working voltage. In repairs, it must always be replaced with the correct type.

D1 is known as the **efficiency diode** or the **recovery diode**. Some manufacturers use the normally reversed biased collector–base junction as a recovery diode.

Figure 17.9 shows the circuit of a line output stage, typical of the type used in a small-screen monochrome receiver. The line scan coils are shown connected in series with the line output transistor. The line output transformer is there to increase the charge on C97 and C99 (flyback tuning capacitors) during the first half of flyback.

EHT (extra high tension)

Additional windings can be added to the transformer to produce such necessities as EHT for the CRT; extra-low or high-tension supply lines or even the CRT heater filament can be operated from one of these windings. The beauty of deriving extra supply lines is that the ripple component

is of high frequency. This makes filtering very easy and cheap. It is far easier to remove ripple from a supply line with a ripple frequency of 15 625 Hz than from one with a frequency of 50 Hz. Small-value capacitors can be used instead of large-value types.

Because d.c. is present on the collector of the transistor the scan coils must be a.c. coupled: that is, via a capacitor. So C98 is there to d.c.-block the supply voltage, and to provide S correction. The screen is wider than it is higher, so the need for S correction is even greater for the line scan than it is for the field scan (more on this subject later).

During the second half of line scan, VT8 (Figure 17.9) is turned on. C98 discharges into the transistor, and current flows upwards through the two coils. Scan ends and VT8 turns off. A very large positive-going pulse is produced because of the transformer. C97 and C99 and any stray capacitance in the circuit now charge up. At this point, current in the scan coils quickly reduces to zero (first half of flyback).

C97 and C99 plus any stray capacitance now discharge into the coils via C98, pushing the current in a downward direction. At this point the current increases to maximum and this forces the beam over to the left-hand side of the screen. The collector voltage attempts to swing negative

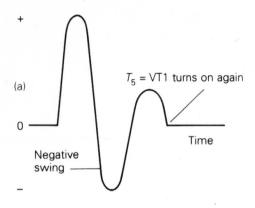

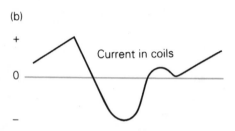

Figure 17.10 The negative swing at the collector of the line output transistor: (a) VT1 collector voltage; (b) possible shape of current in the coils.

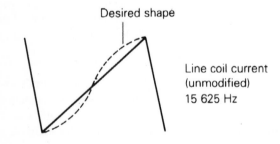

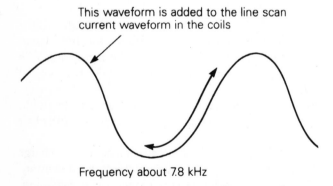

Figure 17.11 Production of the S-corrected scan waveform.

but the diode W17 is biased on. The diode 'damps' the circuit, and the current through the coils is allowed to diminish linearly to zero. This causes the beam to start the first half of the scan.

Without the diode in circuit, the voltage at the collector could produce a negative swing such as that shown in Figure 17.10a. The shape of the current in the coils could cause the beam velocity to vary (Figure 17.10b). This effect could produce vertical **striations** visible on the left-hand side of the screen as vertical light and dark bars: lighter because the beam is travelling slower and darker because the beam is travelling faster.

S correction is cleverly achieved by making the scan coil circuitry resonant during flyback at approximately half line frequency. The value of the d.c.-blocking S correction capacitor is critical. Figure 17.11 shows the basic idea. A waveform of approximately 7.8 kHz is added to the line scan current waveform to achieve the desired shape.

As we have already mentioned, the EHT can also be derived from the line output stage. A step-up overwind can be included as part of the transformer assembly. Small-screen receivers may require 12 kV. For this, the flyback pulses can be simply rectified. Large-screen receivers (colour or monochrome) may require 25 kV. In this case **voltage multipliers** are used. These are separate units, which consist of diodes and capacitors. They multiply the output voltage from the line output transformer to the desired level.

The **diode split transformer** has an integral multiplier. The diodes are built into the transformer casing. Stray capacitance is utilised. This type tends to be more reliable than its earlier counterparts.

Fault diagnosis

Care must be taken when fault finding in the line output stage. In a large-screen colour or monochrome receiver, the voltages can be very high. The transistor collector voltage can exceed 2 kV. Test equipment such as multimeters and oscilloscopes can be permanently damaged if connected incorrectly. Nowadays, the faults that line output stages tend to suffer from are usually that they appear 'dead': that is, no function at all. Remember that if the line output stage is responsible for supplying other things, such as EHT, CRT heaters and extra supply lines, then these extras may also disappear if the line output stage ceases to function. A quick check with a multimeter on line-derived HT lines will confirm the operation of the line output stage.

There are many different types of line output stage:

some have two line output transistors, while thyristors have also been used as the switching device, as well as separate scan and EHT circuits. Some types are tied in with a switch mode power supply.

CHECK YOUR UNDERSTANDING

● The line timebase's main job is to provide sawtooth-shaped currents in the line deflection coils to give horizontal scan. As for the field timebase, S correction must be provided.
● EHT can be developed from the line timebase, plus extra supply lines, which may be needed for the CRT.
● Protection for the CRT circuits is provided by using spark-gap components.

REVISION EXERCISES AND QUESTIONS

1 In all modern television receivers, flywheel synchronisation is used in the line timebase. What is the advantage of using this type of synchronisation?
2 What is the disadvantage of using flywheel synchronisation?

Intercarrier sound

The intercarrier sound signal

The sound IF is at 33.5 MHz. It is frequency modulated. This signal was developed in the mixer circuit in the tuner. It passes down the sound/vision IF amplifier circuit and appears at the vision demodulator.

In transmission the sound carrier was spaced at 6 MHz apart from the vision carrier (or, more exactly, 5.996 MHz). The sound IF and the vision IF mix together in the vision demodulator circuit. A 6 MHz beat signal is produced. Figure 18.1 shows the basic idea.

It can be seen that this difference signal of 6 MHz has on it the two types of modulation, AM and FM. The difference signal has been frequency modulated by the sound signal and amplitude modulated by the vision signal. As the difference signal is supposed to be the new sound IF signal, it is necessary to remove the AM

component. This can be achieved by feeding this signal into a limiter circuit. This will remove the AM component and leave behind the wanted FM component. The new difference signal is known as the **6 MHz intercarrier sound IF.**

The use of intercarrier sound is a **receiver function**: something that is done in the receiver in order to overcome a particular problem. In this case, the problem is **local oscillator tuning drift**. If the oscillator in the tuner drifts off its correct frequency, then the two IF signals will also change in frequency, because they depend on the oscillator in the tuner. If the IFs changed their frequency because of a change in local oscillator they would still be separated by 6 MHz.

What would happen if we did not use the intercarrier principle? To answer this above question we must first look at part of a theoretical television receiver block

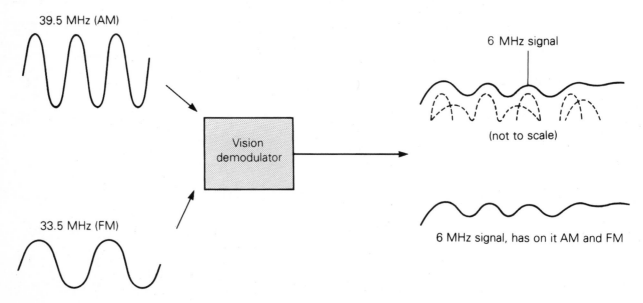

Figure 18.1 Obtaining the 6 MHz intercarrier sound IF.

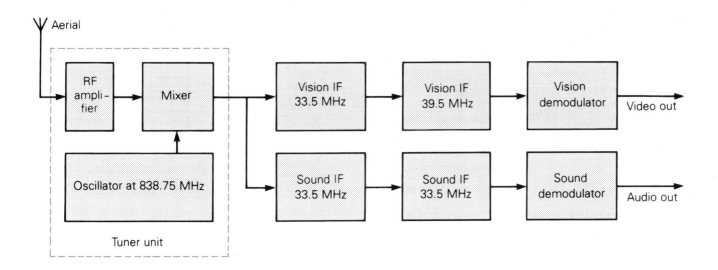

Figure 18.2 Part of the block diagram of a television receiver using separate IF amplifiers.

diagram (Figure 18.2). This receiver has separate sound and vision IF circuits. The sound carrier uses FM. The deviation is ±50 kHz. A bandwidth of about 200 kHz is allowed for the sound channel.

Assume that the television is tuned in to Channel 62. The oscillator in the tuner should be operating at 838.75 MHz. This produces two IF frequencies, one at 39.5 MHz and the other at 33.5 MHz: the vision and sound IFs respectively. Figures 18.3a and b show what the response curves could look like for the two separate IF stages.

If the oscillator drifts upwards in frequency by a very small amount, say 0.02 per cent, this would alter the oscillator to a new frequency of 838.92 MHz. Now, the IF = $f_o - f_s$: therefore

838.92 MHz − 799.25 MHz = 39.67 MHz

The new vision IF would then be about 39.67 MHz. This new frequency would appear on the vision IF response curve as shown in Figure 18.3a. The visible effect of this new IF frequency would be minimal.

The new sound IF is again equal to $f_o - f_s$:

838.92 MHz − 805.25 MHz = 33.67 MHz

The new sound IF would be 33.67 MHz. This new IF would appear on the sound IF response as shown in Figure 18.3b. It does not even fall within the pass band of this curve. The audible effect would be that there would be **no sound**.

Therefore a small percentage drift in the oscillator

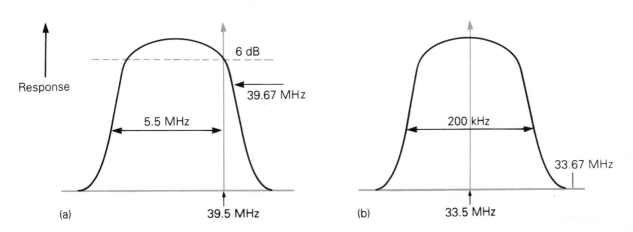

Figure 18.3 The effect of losing the sound carrier because of tuner drift: (a) vision IF response; (b) sound IF response (not to scale).

frequency would first of all cause a loss in the sound IF signal. A receiver of this type would be very difficult to tune. Even with AFC, problems would still be experienced.

For a receiver using the principle of intercarrier sound, any small percentage oscillator drift will not have any effect on the sound signal because the relationship between the sound IF (33.5 MHz) and the vision IF (39.5 MHz) will not alter: there will always be 6 MHz between them.

Intercarrier sound demodulator

Many different methods are used to extract the intercarrier sound signal from the vision demodulator. A tuned circuit may be used after the vision demodulator, or the signal may be taken off at the video driver stage. The last vision IF stage can also be used as the take-off point. Ceramic filters can be used to select the intercarrier signal.

Figure 18.4 shows the circuit diagram of the FM demodulator IC used in the Ferguson 1600 series receiver. A **ceramic filter** (CF1) is used to pick off the intercarrier sound. IF signal, which appears at the output of the vision demodulator.

Ceramic filters are bandpass filters. They can take the place of conventional LC circuits. They are non-tunable (fixed resonant frequency) and cheap to produce. The ceramic filter behaves like a resonant tuned circuit with a high Q factor. The shape and size of the ceramic filter determine the resonant frequency. They are

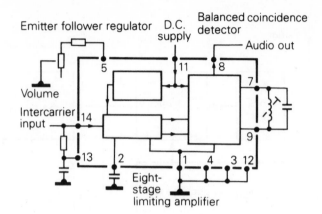

Figure 18.5 The internal functions of the FM demodulator chip IC2.

manufactured in the frequency range of about 20 kHz to 10 MHz.

Figure 18.5 gives a basic idea of the internal functions of this IC. The 6 MHz IF is fed in at pin 14. An eight-stage limiting amplifier is shown here. These stages remove any amplitude modulation that may have been superimposed on the intercarrier signal. Remember that it is the FM signal we want and not any AM signal. This type of FM demodulator is known as a **coincidence detector**.

External LC (L16/C65) circuits cause a phase shift of 90° to occur between two signals (they are said to be in **phase quadrature**). These two signals are themselves derived from the 6 MHz intercarrier signal. At 6 MHz the shift is exactly 90°. As the original modulation causes the frequency of the carrier to alter or deviate, the phase

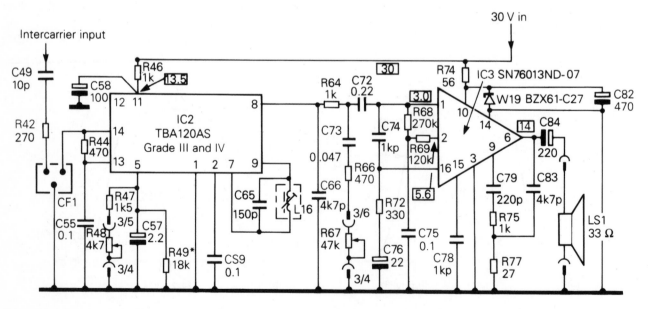

Figure 18.4 The FM demodulator chip IC2 (Ferguson 1600 series). IC3 is the audio output chip.

shift or the amount of coincidence between the two alters. A third signal is generated, which is in the form of a variable pulse width. After suitable filtering the d.c. voltage obtained will depend on the amount of deviation produced by the original modulation: that is, it will swing positive and negative in sympathy with the modulating signal.

A voltage-controlled attenuator (volume control) also exists within IC2. The advantage of using this type of volume control is that the volume control can be positioned anywhere in the TV cabinet, without the fear of the connecting leads picking up any unwanted signals. The leads themselves carry a d.c. voltage and no audio signals. The audio signal is fed out at pin 8 of the chip to the audio preamplifier and power output stage (IC3).

CHECK YOUR UNDERSTANDING

● The vision demodulator demodulates the vision signal, and also passes on the beat frequency 6 MHz intercarrier signal (and the chroma signal in a colour receiver).
● The principle of intercarrier sound is used. This compensates for oscillator tuning drift.

REVISION EXERCISES AND QUESTIONS

1 Intercarrier sound is said to be a 'receiver function': that is, something that is carried out in the receiver itself. What is the main advantage of using this principle?

The monochrome cathode ray tube

Introduction

This chapter describes the operation of a CRT typical of the kind used in a monochrome television receiver. The importance of safe handling is also discussed.

Operation

Figure 19.1 shows a cross-section of a CRT. The internal layout of the gun assembly is also shown. The deflection coils are shown mounted on the tube flare.

The CRT heater (h) raises the temperature of the cathode (k). The cathode is a cylinder made from nickel. The tip of the cathode is coated with a strontium–barium

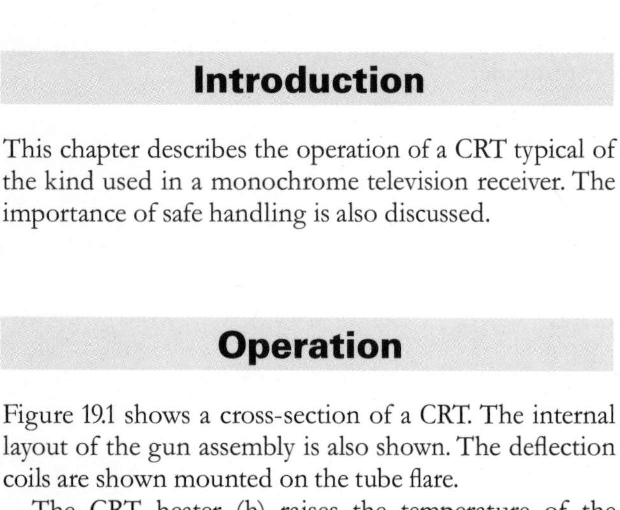

Figure 19.1 Cross-section of a cathode ray tube, showing the basic construction and the internal layout of the gun assembly.

oxide. This coating is rich in free electrons, and when its temperature is raised, electrons are emitted.

The first anode (a1) is at a high positive potential of a few hundred volts. This causes the electrons from the cathode to accelerate towards the anode. The grid (g1) is negative with respect to the cathode. The effect of this is that it controls the flow of electrons through its aperture.

The assembly a2, a3 and a4 forms an electron lens. This lens is similar to an optical double convex lens, in that it brings the divergent electrons to a focus at the screen by virtue of the voltages on these electrodes. The electrodes a2 and a4 are at the same voltage. This is the final anode potential obtained from the EHT connection via the internal graphite coating. A typical figure for the EHT would be about 15 kV.

Anode a3 is at a few hundred volts. This is usually adjustable so that the focus can be varied.

The line and field deflection coils provide horizontal and vertical electromagnetic deflection of the electron beam. The line coils are mounted above and below the tube flare. The field coils are mounted on each side of the flare. The position of the picture on the screen can be adjusted by rotating special **ring magnets** or **shift rings**. These can alter the path of the electron beam prior to the point of beam deflection. The shift rings provide a permanent magnetic field, which permeates through the tube glass into the path of the electron beam. Moving the beam path alters the position of the picture on the screen. When a CRT is manufactured, it is not always possible to ensure that the electron gun assembly is placed centrally in the tube neck. The shift rings are therefore included to allow for this positional tolerance.

The phosphor coating inside the faceplate **fluoresces** (gives off light) when it is bombarded with electrons. The aluminium backing prevents iron burn while at the same time increasing the light output from the front of the tube: that is, it acts as a mirror.

Anodes a1, a2, a3 and a4 are all at positive potentials. It

might be thought that because these electrodes are all at positive potentials they would act as collectors of electrons. This is not so, because they are all cylindrical, with holes at both ends. The force acting on the beam is the same from the inner walls of the cylinder. So the beam travels straight through and out the other side.

The outer edge of the faceplate is surrounded by a mild steel strap about 1 in (25 mm) wide. This is called a **P-band**. The P-band has mounting lugs on each corner to secure the tube in the cabinet. The strap is bound tightly around the glass faceplate. If the tube glass was broken for any reason, without the strap the tube would **implode** (fall in on itself), owing to the very high vacuum in the tube. The compression force of the strap tends to keep the tube in one piece and prevents it from collapsing inwards.

Safe handling of CRTs

There are two important considerations in handling CRTs. One is **mechanical** (tube breakage); the other is **electrical** (the high EHT voltage). Both are dangerous. One can cause injury due to flying glass; the eyes are particularly at risk. The other can result in an electric shock, even when the receiver is off. Colour television tubes are particularly dangerous, as they use an EHT of 25 kV.

How to carry the tube

When you are handling the tube, carry it by the P-band, or with a hand under the faceplate. The faceplate is quite thick: about half an inch at the centre.

> ⚠ Never put pressure on the tube neck. Wear eye safety goggles and gloves when you are handling tubes.

The inner and outer aquadag coatings form two plates of a capacitor, with the glass of the tube acting as the dielectric. The total value of capacitance is about $0.001\ \mu F$. This 'capacitor' actually stores the EHT voltage and provides smoothing. When the receiver is turned off this capacitance can continue to store the EHT for some time. Even after the tube is removed from the cabinet you can still receive a shock from the EHT connection, which may cause you to drop the tube.

Discharging the CRT capacitor

After removing the EHT lead from the tube, join the EHT tube connection to the outer aquadag via a resistor. It is quite easy to make up a special lead and resistor. You will need:

1. a length of EHT lead about 16–17 in (40–43 cm) long;
2. an EHT connector cap – one from an old EHT doubler or tripler unit, for example;
3. a crocodile clip;
4. a small cardboard tube;
5. a 220 kΩ resistor;
6. a small amount of epoxy resin to seal the ends of the tube;

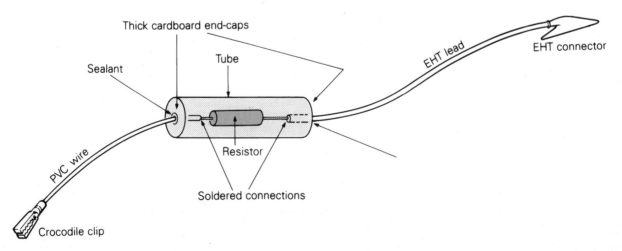

Figure 19.2 A suitable lead for discharging the final anode of a CRT.

7. a small length of PVC-covered wire;
8. solder.

Make the lead up as shown in Figure 19.2. When connected, it will slowly discharge the tube capacitance.

Typical voltages for the monochrome CRT

These vary according to the tube size, but typical values are as listed in Table 19.1.

Table 19.1 Typical voltages for the monochrome CRT

Heater	6.3–12 V d.c. or a.c.
Cathode	100 V (variable)
Grid	0–60 V d.c.
First anode (a1)	300 V d.c.
Third anode (a3) focus	350 V d.c.
Final anode (a2/a4)	12–18 kV

Associated circuits

This term includes the circuitry used for providing the tube operating voltages, and also the means of obtaining the voltages for the first anode, the focus anode and the brightness control circuit. See Figure 19.3.

Heater voltage

The heater can be operated from an a.c. or d.c. supply. Its only purpose is to heat up the cathode. In this case it is operated from a winding on the line output transformer (pins 1 and 8 on the CRT).

Cathode voltage

The cathode (pin 7) voltage is derived from the 180 V supply line. This 180 V line is fed to the video output transistor collector via the collector load resistor. The video signal causes the cathode voltage at the tube to vary between say 170 V (blacker than black level) and say 100 V (peak white level). Blacker than black is caused by the blanking pulse that is added to the video signal (see Chapter 14).

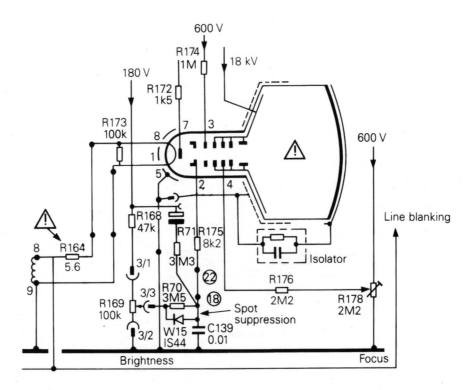

Figure 19.3 The CRT circuitry.

The video signal at the cathode is negative going: that is, blanking pulse uppermost. Peak white gives a beam current of about 150 μA.

The brightness control is connected to the grid circuit (pin 2). The purpose of this control is to set the black level of the picture. If the brightness control is set so that the black level is correct, and the contrast control is set for correct white, then all the shades of grey in between should be correct.

The grid voltage

For this circuit, the grid voltage is variable between 0 and 70 V. This is set by the brightness control R169. If the slider of R169 is moved up, the grid voltage will increase. Let us say that the grid voltage is 30 V, and to produce black level the cathode voltage increases to 150 V. We can then say that the cathode is effectively 120 V positive with respect to the grid, or the grid is 120 V negative with respect to the cathode. At this point the screen is black: that is, black level or no raster.

Now, let us assume that the video signal causes the cathode voltage to reduce to say 100 V: that is, white. We can now say that the cathode-to-grid voltage is 70 V, or the grid-to-cathode voltage is negative by 70 V.

So, for white on the screen there is less reverse bias (more beam current), and for black there is more reverse bias (no beam current): that is, cut off. For white or black on the screen, the grid is always negative with respect to the cathode. The grid must never be allowed to go more positive than the cathode, otherwise damage will occur. The brightness control is used to set the black level of the picture on the screen.

Spot suppression

This circuit consists of C70, R71, R70, W15 and C139. Its purpose is to keep the tube cut off while the cathode cools down. All monochrome receivers have some circuit of this type (see Figure 19.3).

When the receiver is switched off, the cathode remains hot for some time afterwards. Emission from the cathode will still take place. Also, there will be some EHT stored in the tube capacitance. Electrons can still be attracted to the screen. However, the timebases have stopped working. So what will happen is that a spot will be seen in the middle of the screen. In time, the beam will burn off the phosphor at this point, and permanent damage will result.

The spot is suppressed in this circuit by holding off the beam current until the cathode has cooled down. The grid is made to go negative. When the receiver is turned on, C70 charges up through R71, W15 and the brightness control R169 from the 180 V supply line. When C70 has charged up, its negative plate returns to a voltage set by the brightness control.

When the receiver is switched off, the 180 V line collapses very quickly. The bottom plate of the capacitor drops from say 30 V to about −150 V. This negative voltage is applied to the tube grid via R71 and R175, cutting off the beam current. This charge gradually leaks away through R71, R70 and whatever resistance there is between the brightness control wiper and the chassis. The diode W15 is reverse biased and so brings in R70. The time constant of the circuit is about 32 s. This is plenty of time to allow the cathode to cool down and so prevent the CRT from being burned.

Supplies for the first and third anodes

The first and third anodes (pin numbers 3 and 4) are supplied from the 600 V HT rail. The first anode gives beam acceleration. Without it the screen would appear very dark. The third anode is the focus anode. It works in conjunction with the second and fourth anode. This anode is also supplied from the 600 V rail. It is made variable by adjusting R178.

Beam deflection

As already mentioned, the beam is deflected by two pairs of coils: one for the line, and one for the field deflection. The current in the coils is sawtooth in shape, with a small amount of S correction. For a small-screen monochrome portable the currents may peak as high as 1 A. For line scan deflection, the two coils are positioned above and below the tube neck.

The field coils are mounted at each side of the neck. Figure 19.4 shows the basic principle of electromagnetic deflection using the field coils. The electron beam is regarded as a current-carrying conductor in a magnetic field. Applying Fleming's left-hand rule, and using the flow of conventional current, we can determine the direction of movement for the beam. In this case the beam moves down. If the magnetic field changes direction (due to the current) then the beam moves up.

The line coils are mounted at right angles to the field coils.

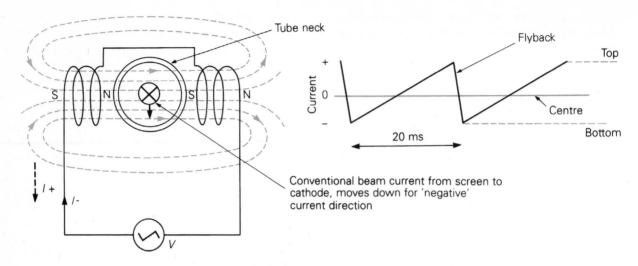

Figure 19.4 Deflection of the beam by a magnetic field.

Table 19.2 Typical faults in a monochrome CRT

Fault	Symptoms
Low emission	Poor picture quality, out of focus, negative picture. This occurs with age.
Open-circuit filament	No picture at all: no emission.
Grid cathode short-circuit	Very bright raster, no video (check with ohm meter).
Detached first anode electrode	Dark picture, highlights, or the bright parts of the picture just visible.
Poor contact between the internal spring and the aquagag coating	Intermittent disturbance of the picture. This effect will depend on brightness level.
Phosphor missing from inside of faceplate	CRT loss of vacuum; i.e. air inside.
Faults with the deflection coils	
One coil open-circuit	Trapezium distortion (see Appendix 3).
Both coils open-circuit	Line or field collapse, depending on which pair.
Short-circuit turns in coil	Again, symptoms depend on which pair. For the line pair, possibly no picture, because of excessive loading in the line output stage. For the field pair, possibly distortion in picture, also overheating of field output transistors.

The isolator unit

The isolator unit shown in Figure 19.3 is designed to discharge the P-band. The P-band may pick up static electricity from the tube area, so it is important that it is discharged via this unit. The P-band cannot, in this type of receiver, be connected directly to the chassis, because it may be accessible from the front of the cabinet. If the receiver chassis became live for any reason then so would the P-band. This is why the P-band is connected to the metal chassis via a high-value resistor. This limits any dangerous currents that might otherwise flow to a safe value.

Monochrome cathode ray tube faults

Typical faults with a CRT of this type are listed in Table 19.2.

ACTIVITIES

▲ Warning. For the following activities, you will be working in a high-voltage, high-current area. Get the assistance of your tutor.

1 For this activity, you will need a monochrome receiver complete with service manual, and a good off-air signal.

Measure the voltages at the following points:

i) the cathode;
ii) the grid;
iii) the first anode;
iv) the focus anode.

Compare the grid and cathode voltages. Note that the grid is less positive than the cathode.

Now obtain a special EHT test meter. Carefully remove the EHT lead from the final anode connector on the tube. Discharge the CRT (see earlier note). Connect the test lead from the EHT meter to the EHT lead. Now try and connect both leads into the final anode socket on the tube. Make sure that the negative lead from the meter is firmly connected to the receiver chassis. Switch on the receiver and read the scale on the EHT meter. Depending on the size of the tube, the measured voltage should be between 9 kV and 18 kV. Note that with an off-air signal and moving scene, the EHT will vary with scene brightness: it will be at its lowest for a bright scene and highest for a dark scene.

2 With the monochrome receiver turned off and the power disconnected, slacken the screw that secures the clamp around the deflection coils. When the coils are free to move, slide them back about 2.5 cm towards the rear of the tube. Now switch on the receiver and note the shape of the picture. What you should now see is that the picture has no corners! The deflection centre has been moved so far back that the electron beam is hitting the inner wall of the tube. This creates a shadow on the picture: that is, a point on the screen where the beam is not reaching.

3 If the field scan coils are wired in parallel (they usually are), try disconnecting one of the coils in a pair. Try the field coils first. If all goes well, you should see a trapezium-shaped raster similar to one of the diagrams in Appendix 3.

4 Obtain a monochrome receiver, and the circuit diagram. Locate the spot suppression circuit. It is usually in the vicinity of the tube. Now find some way of disconnecting it. Ask your tutor for some assistance with this. (Usually disconnecting the large-value capacitor does it.) Note the effect on the tube when the TV is turned off. Do not repeat this exercise too many times, otherwise you will burn the tube!

5 With the monochrome television set up for a stationary picture (test card or similar), place a bar magnet near to but not touching the tube neck, close to the deflection coils. Now describe the effects that you see on the picture. If you are doing this correctly the picture should move position! This proves that the magnetic field from the bar magnet has an effect on the picture position. Do not forget that the job of the shift rings is to provide this field, and so position the picture in the correct place on the screen.

Warning. **Never** do this to a colour tube or you could permanently magnetise the internal assembly. Do not even place a magnet anywhere near the front of the screen. The shadowmask lies behind the glass, and it may become permanently magnetised. Damage will be the result.

CHECK YOUR UNDERSTANDING

- The job of the heater is to heat up the cathode. The cathode emits electrons.
- The grid controls the amount of electrons flowing to the screen.
- The first anode gives primary beam acceleration. The second and third/fourth anodes act as a lens system providing focusing for the beam.
- The EHT voltage can vary, depending on the size of the tube, and can be anything from 10 kV to 18 kV for a monochrome tube.
- The line and field deflection coils are mounted around the tube neck. The line pair are mounted top and bottom and the field pair are mounted on either side.
- Ring magnets are used to position the picture on the screen. They allow for positional tolerances when the gun assembly is fitted into the tube neck.
- The aluminium backing prevents screen burn due to ions. Also, it increases the screen brightness.
- The grid and all of the anodes are made of metal cylinders.
- The P-band is a strap of metal that surrounds the CRT. It prevents implosion in the case of breakage.
- EHT can be dangerous. Gloves should be worn when handling CRTs. Eye goggles must also be worn. Before removing the CRT, it should be discharged.
- The heaters can be d.c. or a.c. controlled. The video amplifier output is connected to the CRT cathode. The video signal modulates the intensity of the beam, and hence the screen brightness.
- Peak white gives maximum beam current and black level gives zero beam current.

- The brightness control is used to set the black level.
- The contrast control is used to set the white level.
- The grid is always more negative than the cathode.
- The switch-off spot suppression circuit prevents screen burn, which otherwise could occur when the receiver is turned off. The grid is held negative until the cathode is allowed to cool down, preventing any more emission.
- The first and third anodes are operated from a high-voltage supply.
- The isolation unit prevents the P-band from acquiring a static charge. Direct connection to the chassis in certain receivers is not possible because of the danger if the receiver chassis becomes live.

REVISION EXERCISES AND QUESTIONS

1 Before a CRT is removed from the cabinet, what is the first job that must be done?
2 What are the possible symptoms on the screen of a monochrome CRT if the first anode electrode became disconnected?
3 A fault develops on a monochrome TV, where the spot suppression circuitry fails. What are the effects on the picture when:
   i) the receiver is turned on and warmed up,
   ii) the receiver is turned off?

Introduction to
the colour receiver

Introduction

The next two chapters deal with chrominance signal processing in the colour receiver decoder. This chapter describes the individual stages involved with this processing, using a block diagram. Chapter 21 discusses fault symptoms and test gear requirements. It then goes on to describe colour CRTs, and compares old and modern types. Typical adjustments are also discussed.

The colour signal

The composite chrominance signal provides information about hue and saturation. It forms part of the composite video signal, and it passes through the colour receiver IF stage with the video signal at a frequency of 35.07 MHz. The chrominance signal subcarrier is centred on a frequency of 4.43 MHz. Suppressed subcarrier modulation is used in the transmitter, so only the chroma sidebands are produced. The highest chroma video frequency is not allowed to extend beyond 1 MHz: therefore 2 MHz is needed for the pass band, ±1 MHz. The first job therefore is to separate the chrominance signal from the luminance signal. The chrominance signal can then be handled separately (decoded) to produce the original R − Y, B − Y and also the missing G − Y. When this has been achieved, we can derive the red, blue and green signals. These signals can then be used to drive the cathodes of the CRT.

We shall now describe the block diagram shown in Figure 20.1. The area within the dashed lines may be part of an integrated circuit, which is now used in many of the modern colour television receivers.

The vision demodulator

Block 1 is the vision demodulator. A switched synchronous demodulator such as that described in Chapter 14 can be used.

The composite video signal and the 6 MHz intercarrier signal are applied to the video amplifier (block 2). Two outputs are supplied from this block: one to the sync separator, and the other to block 3, the band stop filter. The 6 MHz intercarrier sound is taken off between these two blocks. A ceramic filter can be used for this purpose. The band stop filter is used to prevent the 6 MHz intercarrier sound signal from entering the luminance stage. If this did occur then very fine patterning would be seen on the screen.

Between blocks 3 and 4 the chrominance signal is taken from the composite signal. A series-tuned circuit is used to extract the chroma signal. It is then passed on to block 5, the first chroma amplifier. The chroma signal must not be allowed to enter the luminance channel, otherwise excessive dot crawl would be seen on the screen. This would be objectionable to the eye. The effect would be similar to the 4.5 MHz vertical bars on the test card (Appendix 2). In a monochrome receiver no attempt is usually made to reject the subcarrier, because it does not really cause any major disturbance.

The bandwidth of the luminance amplifier (block 4) is 0–5.5 MHz. The tuned circuit mentioned earlier prevents the chrominance signal from getting into the luminance stage. It effectively acts as a notch filter. Figure 20.2 shows the effective frequency response curve of this luminance amplifier. It should be as flat as possible over the range of d.c. to 5.5 MHz. The tuned circuit mentioned previously not only takes the chroma signal away from the luminance stage but also acts to effectively cut a notch out of this

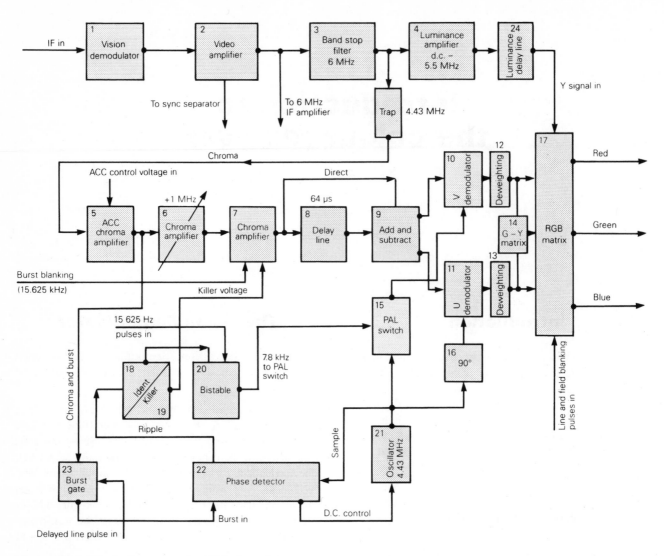

Figure 20.1 Block diagram of a PAL decoder.

response curve, making the gain at this frequency low.

Block 24 is the luminance delay line. It prevents the luminance signal from arriving at the CRT cathode before the chrominance signal. A typical delay time for this component would be somewhere between 0.5 and 1 μs (see Chapter 13). The luminance signal now enters the RGB matrix circuit (block 17).

Block 5 is a chroma amplifier, the gain of which is controlled by a d.c. voltage. Under poor reception conditions, the chrominance signal may vary in amplitude. The job of this stage is to try and keep the amplitude reasonably constant. It works by measuring the size of the burst signal (not shown). From this, a d.c. voltage is derived. The automatic chrominance control (ACC) stage is similar to an AGC circuit in a radio or television receiver.

Block 6 adjusts the balance between the luminance and the chrominance signal: that is, the saturation or the

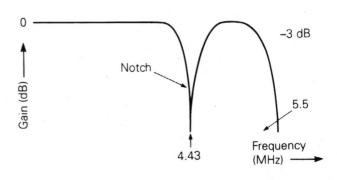

Figure 20.2 The result of using a notch filter on the bandwidth of the luminance amplifier.

amount of colour in the scene can be adjusted by means of the customer colour control. Varying the amplitude of the chroma signal exiting from this block will alter the value of the R − Y and B − Y signals, thereby varying the saturation level. The control is usually adjusted for natural flesh tones.

Block 7 does two things. First, it prevents the burst signal from entering the chrominance channel and, second, it turns off the chrominance amplifier for a monochrome transmission (colour killer). The burst signal must not be allowed to enter the chrominance demodulators, otherwise it might be demodulated. This could cause false colours to appear on the left of the screen. This suppression of the burst signal is known as **burst blanking**. This is achieved as follows.

A delayed line flyback pulse is fed into block 7. This pulse corresponds to the same position as the back porch (hence the reason for its being delayed). The pulse is made to turn off the chroma amplifier during this period of time, just like closing a gate. Therefore, the burst signal does not pass to the output. This chroma amplifier may also be known as the **delay line driver amplifier**, because it makes up for the loss in the chroma delay line that follows it.

For a monochrome transmission the chrominance amplifier (block 7) must be turned off. The reason for this is noise: not the kind of noise that can be heard, but visible noise that can be noticed on the screen. If the chrominance amplifier was not turned off, it would allow this noise to be passed straight through into the chrominance demodulators. The effect would be random coloured speckles on the screen. This would impair the monochrome picture.

A d.c. control voltage is fed in to kill the operation of the chroma amplifier. This is known as the **colour killer** action. The d.c. control voltage is derived from the ident signal (block 19) (described below).

The chrominance delay line

Block 8 is the chrominance delay line. The job of the chrominance delay line is to store or delay one complete line of chroma signal. The delay time is approximately $64\,\mu s$.

The composite chroma signal enters block 8. As one line of chroma, say line $n + 1$, is entering the block, line n chroma is leaving the block. The signal path through the delay line is known as the **delayed path**. There is also the direct path. Both paths are shown more clearly in Figure 20.3. Block 9 is an adder and subtracter circuit.

When U + V (the composite chroma) is at point X, then at the same time U − V (due to line alternation and delay) is at point Y. The delay line delays signals by about $64\,\mu s$. U + V and U − V are **added** together in the adder. Then on the adder's output, 2U will appear:

$$(U + V) + (U − V) = 2U$$

U + V and U − V are **subtracted** in the subtractor, giving an output of 2V:

$$(U + V) − (U − V) = 2V$$

A typical adder and subtractor circuit usually consists of a centre-tapped coil attached to the output of the delay line. The additions and subtractions take place inside the windings. It can be shown that for every case of polarity change, for the U or for the V signal, separation takes place. It will be recalled (Figures 124a and b) that the U and V signals can change polarity with picture content. In addition, the V signal polarity is alternated line by line.

It can be seen from this that the delay line and matrix circuit effectively separates the V and the U signals. They end up being twice as big as before, but this does not matter.

How does this delay line and matrix circuit cancel phase

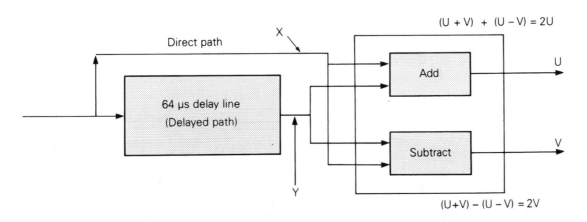

Figure 20.3 Block diagram of the delay line and matrix.

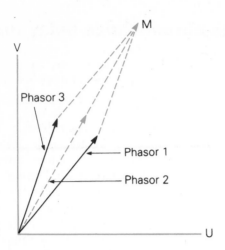

Figure 20.4 Addition of two chroma lines. Phasor 1 is added to phasor 3. The resultant, phasor 2, is nearly twice as big, but the correct colour.

errors? To answer this, we need to look at Figure 20.4. Phasor 1 is the NTSC or unswitched chroma line. An error is shown, because it is lagging phasor 2. Phasor 2 is the correct position for the colour magenta. Phasor 3 is the position for the switched chroma line (PAL). It is now showing a lead on where it should be. Phasors 1 and 3 are added together to produce phasor 2. Now, phasor 2 is the correct colour. The saturation of this colour is slightly less than before, but this is not too important. A slight reduction in saturation is not as noticeable as a change in colour produced by this phase error.

The device that produces the delay of about 64 μs is made of a special glass. The chroma signal is converted by a transducer into mechanical vibrations. These vibrations travel through the glass and are reflected several times. They are finally picked up by a second transducer and then converted back to an electrical signal. For complete cancellation, the delay time is exactly 63.943 μs.

The synchronous demodulators

The V and the U signals are now passed to the V and the U demodulators (blocks 10 and 11). They are of the synchronous type. Normal envelope detection cannot be used with signals that do not possess a carrier; the detector/demodulator would produce the wrong shape at its output. The two chroma signals consist of sidebands only, and therefore a special active demodulator must be used. Also, a new carrier signal must be generated, which must be at the same frequency and phase as the one that was suppressed in the encoder.

The oscillator in block 21 acts as the missing carrier, and it is used to sample the V and the U signals every cycle.

The V and the U signals are also 90° apart from each other. Remember that this was done deliberately so that the two signals could be separated. Therefore the reference oscillator's feed to one of the demodulators is phase shifted by 90°. Block 16 shows this phase shifter included in the feed to the U demodulator.

Also, for PAL, the V chrominance signal is inverted on every line. Facilities must be included to cope with this change. Block 15 is known as the **PAL switch** or **vertical axis switch**. Its job is to change the phase of the reference oscillator feed to the V demodulator every line. If this were not done, then incorrect colours would result for the PAL chroma line. Figure 20.5 illustrates the basic operation of the V demodulator.

Figures 20.5a and b show two successive lines of chroma signal: the NTSC line and the PAL line. The V chroma signal is applied to the electronic switch, as shown in Figure 20.5d. The electronic switch is made to close when the reference signal reaches its positive peaks, at points a, b, c, d, e, f, g, h, ... etc. The switch is open for all negative peaks of the reference signal. Voltage points that are between the two lines of chroma signal are not important because the chroma signal is at zero level. Note that the burst signal has been removed by the burst-blanking stage.

The idea is that when the switch closes, the half cycles of chroma signal are transferred to the filter capacitor. The peaks of the sine wave chroma signal charge it up. Positive or negative half cycles of the chrominance signal are allowed through to charge the filter capacitor. The output voltage from the filter is shown in Figure 20.5e. As can be seen, the voltage charge can swing positive or negative. The basic outline of this voltage is recognisable as the demodulated V chroma signal.

What happens when the V chroma signal changes polarity, as it does in the PAL system? Figure 20.5b shows this. Figure 20.5c shows that the reference signal from the oscillator has also changed phase (point X). The PAL switch causes this.

The phase inversion of the reference signal must follow the phase inversion of the V chroma signal. If this did not occur, the demodulator would produce the wrong-shaped voltage at the filter. The viewable result would be incorrect colours.

Figure 20.6 shows the block diagram of the U demodulator. The process is identical to that of Figure 20.5 except that the U reference signal does not change phase: it is of the same phase for all chroma lines. The U reference signal lags the V reference signal by 90°. For simplicity, this has not been shown. For both Figures 20.5 and 20.6, the individual cycles of subcarrier and reference are not drawn to scale, so that it is easier to

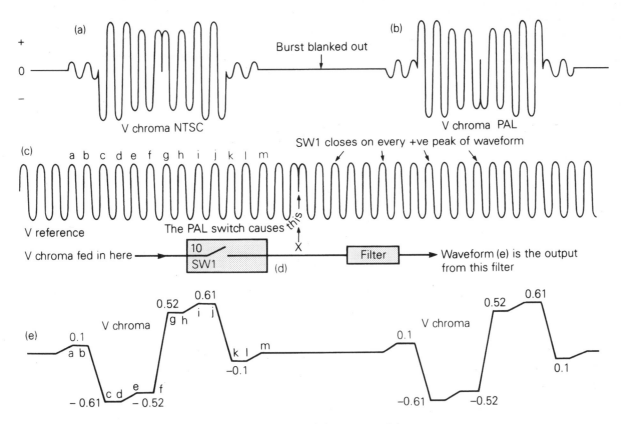

Figure 20.5 The switched demodulator (synchronous) (not to scale).

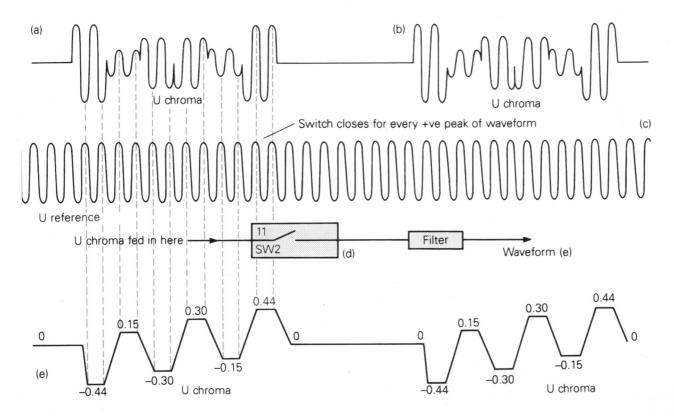

Figure 20.6 The U demodulator and waveforms (not to scale).

visualise what actually happens in a demodulator of this type.

Diodes are used to perform the switching action. The V and the U reference signals have to be relatively large compared with the maximum amplitude of the chroma signal, otherwise the diodes used in the switch will not turn on.

In modern receivers the two demodulators are part of an integrated circuit, with many of the other functions, such as the PAL switch, the reference oscillator, the chroma amplifiers and the RGB matrix. In the latest generation of colour receivers even the PAL delay line is included in an IC package.

Deweighting

Blocks 12 and 13 perform the deweighting of the two chroma signals V and U.

In the encoder the two chroma signals R − Y and B − Y were made smaller in order to overcome the problem of overmodulating the main vision carrier. They were renamed V and U. Now is the time to bring them back to their correct values. The two chrominance signals must now be fed into two amplifiers, whose gain values must equal the **reciprocal** of the values used for attenuation. For V, this is 1/0.877 = R − Y. This gives a gain of 1.140. For the U signal the amplifier gain would be 1/0.493 = 2.03. Another way of putting this is to say that the ratio of the U weighting factor to the V weighting factor is 0.493/0.877 = 0.56. So therefore the ratio of the B − Y amplifier gain to the R − Y amplifier gain must be the inversion of this ratio: that is, 0.877/0.493 = 1.78. This means that the gain of the amplifier used for deweighting the U chroma signal must be 1.78 times that of the amplifier used for deweighting the V chroma signal.

Figure 20.7 shows the two de-weighting amplifiers used for this process.

The G − Y signal

The G − Y signal is not transmitted with the two colour difference signals R − Y and B − Y (V and U). As we know the values of the luminance, the R − Y and the B − Y signal, we can determine the value of the third missing colour difference signal, the G − Y signal.

The formula for determining the G − Y signal is

$$G - Y = -0.51(R - Y) - 0.19(B - Y)$$

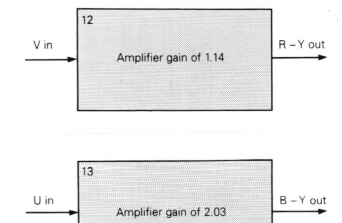

Figure 20.7 The deweighting process.

This expression tells us that we can obtain the G − Y signal by taking about one-half of the R − Y signal and about two-tenths of the B − Y signal. If they are negative going then the resultant G − Y signal will be positive going. Of the three colour difference signals, the G − Y is the best one to be left out from the transmission. It tends to be small in value, and so therefore would suffer from a poorer signal-to-noise ratio compared with the R − Y and B − Y signals (see block 14).

Figure 20.8 shows a resistor matrix circuit used for extracting the G − Y signal.

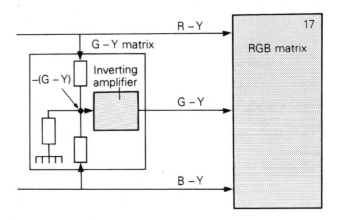

Figure 20.8 The G − Y matrix.

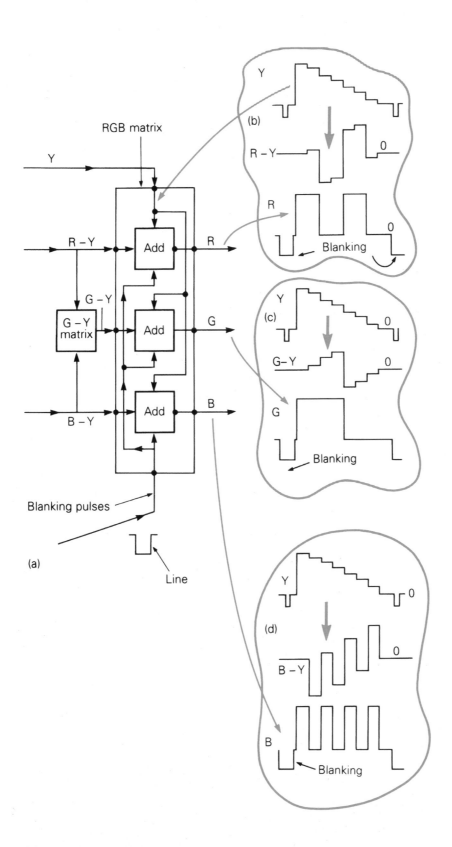

Figure 20.9 The RGB matrix: waveforms for colour bars.

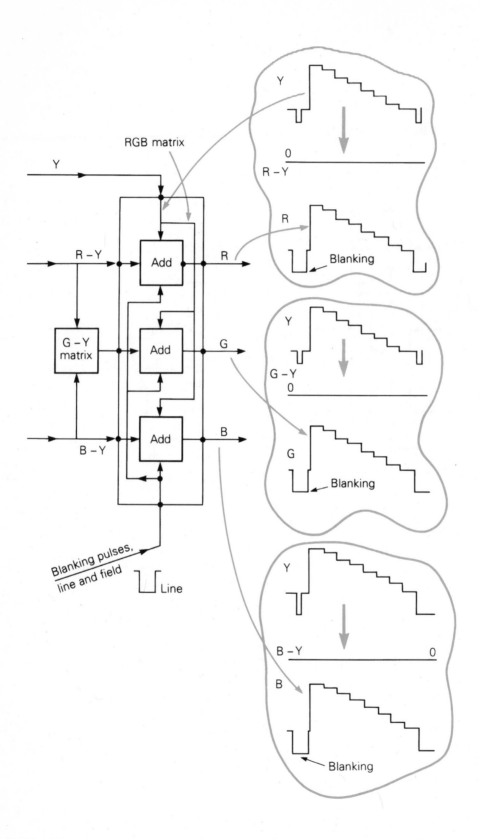

Figure 20.10 The RGB matrix: waveforms for monochrome bars.

The RGB matrix

The matrix shown in Figure 20.9a shows how the luminance signal is added to each of the colour difference signals. The process is as follows.

The luminance signal is added to the $R - Y$ signal, $Y + (R - Y)$. This equals red, R (Figure 20.9b). For green, the luminance signal is added to the $G - Y$ signal, $Y + (G - Y)$ (Figure 20.9c). This equals green, G. For blue, the luminance signal is added to the $B - Y$ signal, $Y + (B - Y)$. This equals blue, B (Figure 20.9d). Blanking pulses (line and field) are also fed in at this point, and are also added to the three signals.

For monochrome pictures, Figure 20.10 shows the $R - Y$, $B - Y$ and $G - Y$ signals at zero. Therefore the only signals emerging from the R, G and B outputs are the luminance signals.

> The R G B matrix produces the colour drive signals. These signals are used to drive the cathodes of the picture tube.

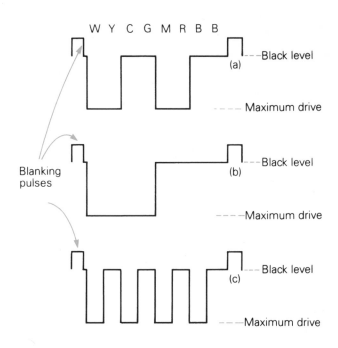

Figure 20.11 Drive waveforms for the three cathodes: (a) red; (b) green; (c) blue.

The R, G and B output stages

In early colour receivers the luminance signal was applied to all three cathodes simultaneously, and the colour difference signals were fed to each of the tube grids: that is, the tube acted as the final matrix. Pictures produced by this method were of high quality. Later on, developments in technology produced transistors that were capable of working with high collector voltages. This meant that receiver circuitry became simpler and less costly. Further developments in IC technology also reduced the component count of the receiver. In modern receivers the trend is to use one, two or even three transistors for each of the colour drive amplifiers.

The job of the RGB output stage is to provide enough signal drive to modulate the beams of the three electron guns. The amount of drive would be in the order of 100–180 volts peak to peak; the actual figure depends on the tube size. The drive signals to each of the cathodes may be adjustable to compensate for different phosphor efficiencies. The bandwidth of each of the stages must be in the order of d.c. to 5.5 MHz.

Figure 20.12 overleaf shows the circuit diagram of an RGB output stage, as used in the Ferguson TX90 series. TR601, TR602 and TR603 are the green, blue and red output transistors respectively. Each transistor is allocated its own particular cathode. The transistors are all working in class A. HF compensation is provided by C602, C603 and C604. Potentiometers RV624, R625 and RV626 are used for setting the black level of each gun. They should all be adjusted so that the three guns cut off together. Spark gaps SG601, 602 and 603 protect the transistors in the event of the CRT flashing over. The RGB drive signals at the bases of the transistors are inverted and amplified. This gives the correct drive conditions for the tube.

Figures 20.11a, b and c show the drive waveforms at each cathode for one line of colour bar signal.

The reference oscillator

Block 21 is the reference oscillator. Its job is to switch the V and the U demodulator circuits so that the V and the U chroma signals can be demodulated. The frequency and phase of this oscillator must stay within a very close tolerance. The frequency is 4.433 618 75 MHz ±10 Hz. The oscillator is crystal controlled by a comparator circuit, which keeps the frequency correct. This comparator circuit (block 22) is also known as a **phase detector**.

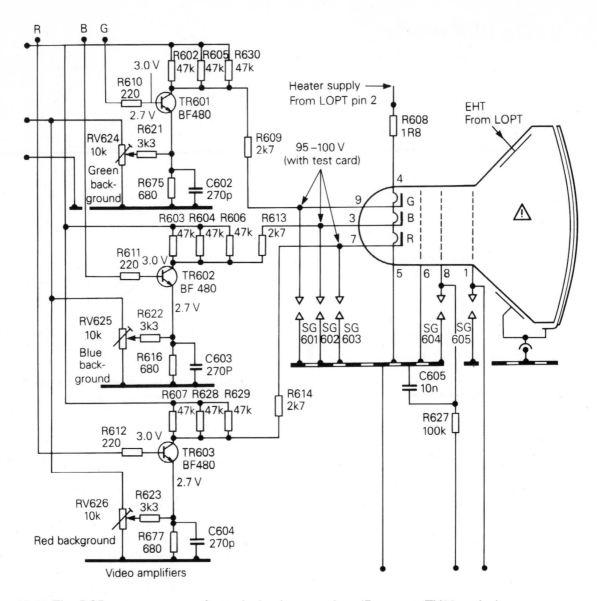

Figure 20.12 The RGB output stages of a typical colour receiver (Ferguson TX90 series).

The transmitted burst signal is gated out by the burst gate (block 23). A delayed line pulse is fed into the gate, together with composite chroma. The gate opens and allows the burst signal to be applied to the phase detector (block 22).

Remember that the burst is swinging in phase: 135° on the normal (or NTSC) chroma line and 225° on the inverted (or PAL) chroma line. But the average phase of the burst is along the −U axis: that is, 180°. The oscillator is designed to lock at 90° lagging behind the average phase of the burst: that is, at 90°, or along the V axis. This is the correct locking position for the oscillator. Figure 20.13 shows the two phases. The phase detector has two

inputs: the burst signal itself and a sample of the oscillator output. The d.c. control voltage is responsible for keeping the frequency of the oscillator correct to within a very small tolerance.

Block 15 is the PAL or vertical axis switch. This alters the phase of the 4.43 MHz reference signal. Figure 20.14 shows a small portion of the reference signal. The phase change of the reference signal is made to occur in the same time frame as the trailing edge of the line sync pulse. This is because the PAL switch is operated by the bistable multivibrator circuit (block 20). A bistable circuit needs to be triggered by an external signal, and always divides by two. The external triggering signal comes from the

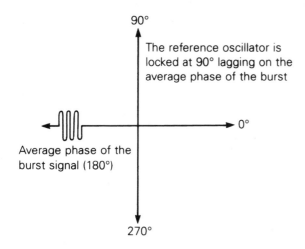

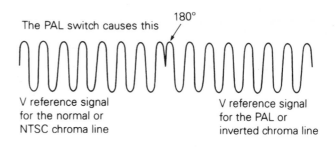

Figure 20.14 A small portion of the 4.43 MHz reference signal.

Figure 20.13 The oscillator is locked at 90° behind the burst.

line timebase: that is, line flyback pulses, which are at a frequency of 15 625 Hz. Therefore the bistable circuit runs at 7.812 kHz or, more accurately, 7.8125 kHz.

Figure 20.15 shows the relationship between the switched reference signal, the square wave signal developed from the bistable circuit, and the line pulses that are used to trigger the bistable. The relationship between the frequency of the 4.43 MHz V reference signal and the 7.812 kHz bistable signal is not strictly correct in this diagram, but it is shown like this to help

in understanding what actually happens. There are in fact 283.75 cycles of V reference in every 64 μs.

In some earlier colour receivers switching of the V chroma signal took place instead of the V reference signal. This led to colour distortion, so it was decided that it was better to switch the V reference rather than the V chroma. It is much easier to switch a simple sine wave than a complex chroma signal: therefore most colour receivers switch the V reference signal.

Remember what is actually happening. The PAL switch alters the phase of the V reference signal fed to the V demodulator: 0° on the NTSC chroma line and 180° on the PAL chroma line. On the next chroma line, it alters it back to 0° again, and so on. The PAL switch is

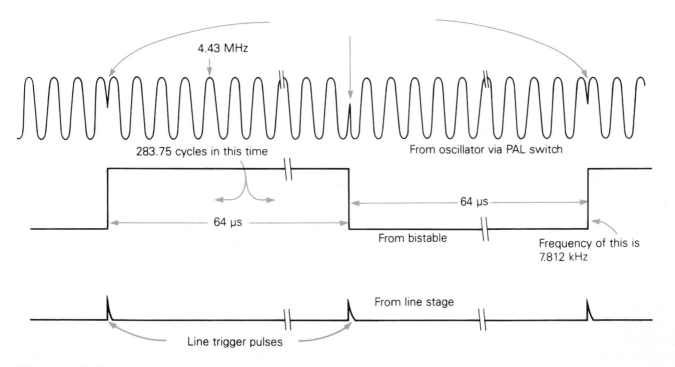

Figure 20.15 Relationship between the bistable signal and the line pulse switching circuit.

operated by the bistable circuit (block 20). Line flyback pulses trigger the bistable, and this causes it to operate at 7.812 kHz. It is possible for the bistable circuit to start up in the wrong phase (that is, 180° out of phase) when the receiver is turned on, or when changing channels. This would cause the V reference signal for the normal chroma line to be inverted when, quite clearly, it should not be. This would cause incorrect colours to appear on the screen for every picture line. Clearly, the bistable circuit needs to be synchronised. This is the function of the 7.812 kHz signal input to the bistable circuit.

The **ident circuit** in block 18 is responsible for synchronising the bistable multivibrator. It is called an ident circuit because it gives out an identification signal, the polarity of which tells the bistable which chroma line is being received: an NTSC chroma line or a PAL chroma line. The frequency of this ident signal is 7.812 kHz.

The ident signal is produced from a tuned amplifier. Its input signal originates from the phase detector circuit (block 22). In the phase detector a sample of 4.43 MHz from the oscillator (block 21) is mixed with the swinging burst, and from this is produced a voltage that contains a ripple of 7.812 kHz. This signal is used to switch the tuned amplifier on and off. From this tuned amplifier a large-amplitude sine wave or (if the bistable prefers) a square wave signal is produced. The polarity of this signal will indicate or identify the presence of an NTSC chroma line or a PAL chroma line.

Block 19 is tied in with block 18. Its job is to develop a d.c. control voltage that can be used to control the third chroma amplifier. In the event of a monochrome transmission the burst signal is turned off. When this happens, the ripple voltage to the ident stage is removed. No ripple means no ident signal, and so the d.c. voltage or killer voltage can be used to switch off the chroma amplifier. This is done to prevent unwanted noise from finding its way to the output and to hence the CRT cathodes.

The luminance delay line

The luminance delay line (block 24) is included simply to delay the luminance signal. It takes much longer to process the chrominance signal than to process the luminance signal. The chrominance channel is also of narrow bandwidth compared with the bandwidth of the luminance channel. This fact also increases the time difference between the two signals. It is therefore important to ensure that the two signals arrive at the output of the RGB matrix, or the CRT cathodes, at the same time.

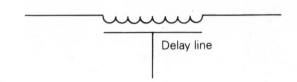

Figure 20.16 The circuit symbol for a luminance delay line.

The method used is to delay the luminance signal by a small amount of time so that the two signals do arrive together. The amount of time depends on the amount of circuitry involved, but a typical figure for this delay would be somewhere between 0.5 and 1 μs (or 500–1000 nanoseconds).

A special coil of wire is used. It has along its length an internal strip of copper, which is connected to the chassis. The effect of this copper strip is to add capacitance to the line, and as a result the signal is effectively slowed down. The signal appears at the output of the line, delayed in time to coincide with the chrominance signal. The circuit symbol of a delay line is shown in Figure 20.16.

Conclusion

This concludes the basic description of the PAL decoder that is used in all PAL receivers. You will gain more experience by looking at manufacturers' service manuals. Remember that the decoder circuits shown in some of these manuals are of a simplified form; the main circuitry is inside an integrated circuit.

ACTIVITY

This activity demonstrates the need for the luminance delay line. You will need a colour receiver, a soldering iron, a simple calculator and a ruler.

Measure the width of the front of the television screen. For a 22 in (56 cm) tube/screen this is the diagonal measurement of the front: that is, from one corner to the other. For a 22 in tube the front width should measure about 17.6 in (44.7 cm). The video signal's active part (the part you can see on the screen) lasts for about 52 μs. Therefore it is necessary to determine what fractional measurement of the screen is equal to about 0.6 μs, which is typical for the delay time of the luminance delay line:

$$\frac{0.6\,\mu s}{52\,\mu s} \times 17.6\,\text{in}$$

$$= 0.2\,\text{in}\ (0.51\,\text{cm})$$

Carefully remove the luminance delay line coil. While it is out of circuit, switch on the receiver and tune to an off-air transmission. Note the effect produced on the screen. What you should see is that the chrominance information is still reaching the cathodes, but the luminance signal is not. Patches of colour should be noticeable, with very poor low-frequency definition. This symptom is described as **no luminance**.

Solder in a wire link to replace the luminance coil. When all is safe, switch on and tune to a normal picture. You should see a 'ghost image'. The chrominance signal should be positioned slightly to the right of the luminance signal by about 0.51 cm. Confirm this measurement by using your ruler. This means that the luminance signal is arriving on the screen before the chrominance signal.

This activity has assumed a delay line of 0.6 μs. Longer delay times will give greater distances between the luminance and the chrominance pictures. Pictures shown on larger tubes will enhance this effect.

CHECK YOUR UNDERSTANDING

- The colour signal is known as the chrominance signal. The chrominance signal gives information about the colour in a scene. On a monochrome signal, the chrominance signal falls to zero. The chrominance signal is usually taken off at or about the vision detector for separate processing.
- A notch filter is included in the luminance path to prevent the chrominance signal from producing a dot pattern on the screen. This reduces the gain of the luminance amplifier to a very low level at this frequency (4.433 MHz).
- The chrominance delay line and matrix circuit separate the composite chrominance signals into V and U. They also cancel out phase errors, which may occur because of the effect of differential phase distortion. Incorrect colour produced by a change of scene is also averaged out.
- Synchronous demodulators are used to demodulate the V and U signals. They each require reference inputs from a 4.43 MHz oscillator. The two reference signals should be 90° apart.
- The two demodulated chrominance signals V and U are deweighted to produce R − Y and B − Y. This must be done otherwise incorrect colours will be seen.
- The G − Y signal is derived by taking about 50% of

the −(R − Y) and about 20% of the −(B − Y) signal. This is done in the G−Y matrix.
- The RGB matrix adds together: the R − Y to the Y producing R; the B − Y to the Y producing B; the G − Y to the Y producing G. Blanking pulses may also be added at this stage. R, G and B appear at the matrix output.
- The R, G and B amplifiers are responsible for driving the three cathodes of the CRT. Their bandwidths must be large: from d.c. to 5.5 MHz in the case of luminance-only signals (that is, monochrome transmissions).
- The reference oscillator must work at the same frequency and phase as the original oscillator in the encoder. The burst signal is transmitted so that this may be achieved.
- The oscillator's job is to demodulate the two chroma signals. It does this by switching the two demodulators on and off.
- The burst sits on the back porch. About ten cycles are transmitted after every line. The burst signal swings in phase: 135° on a normal (NTSC) chroma line, and 225° on an inverted chroma (PAL) line. The average phase is along the −U axis.
- The reference oscillator locks at 90° behind this average phase: that is, along the +V axis.
- The burst must not be allowed to enter the chroma channel. Burst blanking is used to prevent this.
- The chroma signal must not enter the phase detector. Burst gating prevents this from occurring. Line flyback pulses are used to gate this circuit.
- The sample from the oscillator and the burst are compared in the phase detector. This circuit produces an error signal, which keeps the oscillator running at the correct frequency and phase.
- The phase detector also produces a ripple signal at 7.8 kHz. This is used to operate the ident amplifier. As the name suggests, this circuit produces an identification signal, which is responsible for synchronising the bistable circuit. The bistable is also triggered from line pulses. The operating frequency is 7.8 kHz.
- The PAL switch is responsible for changing the phase of the V reference signal for the PAL chroma line: 0° for an NTSC line and 180° for a PAL chroma line.
- The ident stage usually operates a colour killer circuit. From this, a d.c. bias is produced that will enable or inhibit one of the two or three chroma amplifiers. For a monochrome transmission, the chroma stages must be shut down, otherwise noise will appear on the screen.
- Automatic chrominance control (ACC) is used to maintain the same level in the event of varying signal conditions.
- The luminance delay line slows down the luminance signal to ensure that it arrives at the screen at the same time as the chrominance signal.. A typical delay might be somewhere between 0.6 and 1 μs (600–1000 ns).

1 What is the main difference between a simple PAL receiver and a PAL D receiver?
2 Referring to question 1, how are the phase errors cancelled out?
3 What is the name of the special circuit that is included as part of the luminance amplifier in a colour TV? (This circuit prevents dot pattern interference from appearing on the screen).
4 Why is it necessary to include a luminance delay line in the signal path of the luminance signal?
5 With reference to question 4, what would be the visible effect on the picture if this delay did not occur? (Assume a delay time of 0.9 μs and a tube size of 26 in.)

Other subjects

Fault symptoms

In this section we look at what can go wrong within a PAL decoder, and describe any symptoms that may appear on the CRT screen.

Modern decoder circuits are quite reliable, because most of the decoder circuitry is included as part of an IC package. However, there are still plenty of discrete component decoders around.

First of all, you have to identify the type of fault: no colour, wrong colours, unlocked colours or weak colour. Is it possible to obtain a good monochrome picture on the screen? Is the picture definition good? Is there an overall colour cast? Is it just the case that the receiver is mistuned? Check the signal strength. Is it of good quality?

Even if the line hold control needs resetting periodically then this could cause the colour to drop out of the picture. The burst gate circuit relies on the correct timing of this line flyback pulse. If the timing is wrong, the burst will not be gated, and the ident signal will disappear. This will causes the colour killer circuit to operate, killing the chroma signal.

Test equipment

For decoder servicing you will need a good general-purpose **oscilloscope**. It should have a large bandwidth: at least 10 MHz, but 20 MHz would be better. It also has to be sensitive: a maximum sensitivity of at least 10 mV/division. This is because you will also need a high-impedance oscilloscope **probe**. This type of probe attenuates the signal under examination by a factor of 10, and the oscilloscope has to make up for this loss in signal.

You will need a **pattern generator** capable of supplying full colour bars. You cannot trace chrominance signals if the picture is constantly moving. The colour bar generator will display the three primary and three complementary colours, plus white and black.

You will need a **frequency counter** or **frequency meter**. This can have many uses: for example, checking the frequency of the reference oscillator.

You will also need a high-impedance **multimeter**.

No colour

The first thing to do is find a way to override the **colour killer** stage. You will need different methods for different models. Turn the colour control to maximum. If, after overriding the killer circuit, the colour has not returned to the screen, then using the oscilloscope start to trace the chroma signal. Start in the chroma amplifiers. Check the **reference oscillator** (it may have stopped). Remember: no oscillator means no demodulation and also no ident signal.

Unsynchronised colours or rainbowing

You will recognise this effect as soon as you see it: it consists of rows of colours running up or down the screen. Is the oscillator frequency correct? Try adjusting the **frequency control**. A fault in the **phase detector** could also cause this effect. Check also the **burst gate** circuit. The **timing** could be incorrect. Even simple adjustments could cure this symptom.

The colour returns
If normal colour returns to the screen then the fault is most likely to be in the **colour killer** circuit itself. Also check the **ident amplifier** circuit. The ident signal may not be big enough to provide a large d.c. unkilling voltage.

The **PAL switch** may not be working in the correct phase, or may be 180° out of phase. Some of the colours appear to be reversed or in the wrong place. Also, there appear to be some different colours.

What is happening is that the normal line of V chroma signal is being reversed instead of the PAL chroma line. As a result, red appears to be green; green appears to be red; yellow is greenish yellow; blue shows hardly any effect; magenta is a bluish cyan; and cyan is a weak magenta.

This fault may be intermittent; changing channels or even turning the receiver off and on may clear this peculiar effect. The bistable circuit has no direction from the ident stage. This could mean that the ident is weak or is even missing. A check with an oscilloscope will soon reveal the source of the trouble.

The **PAL switch** may be completely stopped. The bistable drives the PAL switch, so it is in this area that you should begin your investigation. Close inspection of the screen will reveal severe Hanover bands. At normal viewing distance you will see peculiar shades of colour.

This means that all the odd lines of chroma signal (1, 3, 5, ... etc.) will be correct. Inspection with a magnifying glass will reveal that two lines of chroma signal will be correct. Why two lines? Remember the two interlaced fields. All odd-numbered lines from both fields will have correct chroma lines. The even-numbered lines will be of incorrect colours, and again there will be two of them. Therefore two correct chroma lines will be followed by two incorrect chroma lines, and so on.

What will the picture look like from a distance? The red bar will be greenish yellow. The magenta bar will be a mixture of cyan and magenta. The blue bar will be slightly magenta. The yellow bar will end up slightly greenish. The green bar will be a mixture of green and orange.

Colour casts

Complementary colour casts

This would indicate that one of the primary colours is missing. If the overall cast is yellow, then this means that the blue information is missing from the screen. The fault could be in the blue video output stage. A fault with one of the transistors or associated components could cause this symptom. Again, a quick check with the oscilloscope will reveal the source of the trouble. The same applies to the red or the green video output stages. A lack of green will give an overall magenta cast; and a lack of red will give an overall cyan cast.

The fault area could be around the V or the U

demodulator circuits. Any fault that prevents one of the primary colour signals from appearing on the screen will give a symptom that will be the result of the other two. For example, green missing will produce magenta (red and blue).

Colour difference signal missing

This is a rare condition for modern receivers, as most of the decoder is sited within an integrated circuit.

If the R − Y signal is missing there will be a lot of blue and green in the picture. People's faces will appear dark green. The colour bars will be white, green, cyan, green, blue, black, blue and black. For a missing B − Y signal, the colour bars will be white, yellow, greenish cyan, green, a very red magenta, very dark blue, nearly black, and black. For a missing G − Y signal the colour bars will be white, reddish yellow, bluish cyan, dark green, pale magenta, orangeish red, mostly blue with green, and black.

These symptoms are only a guide to what you might see on the screen. It also assumes that in each case the colour difference signal is lost after the G−Y matrix circuit. If a colour difference signal is lost before the G−Y matrix (in one of the demodulators or even in the deweighting process), then you would see different symptoms.

Colour television tubes

History

The first shadowmask CRT was made by Dr Harold B. Law in 1949, while working for RCA in the USA. Picture quality was very poor. The picture size was in the order of 4 in (10.1 cm). The basic design was comparable to today's tubes: it had three guns, a shadowmask and a three-colour phosphor screen. The guns were mounted at 120° apart from each other.

In 1954, tubes were manufactured with 70° deflection angles. They were also circular (remember some of the old American movies?), and 21 in (53.3 cm) in size. Later, deflection angles increased to 90°. The tubes were more rectangular than round. The size also increased, to about 25 in (63.5 cm).

In Japan in the late 1960s Sony produced a type of tube called a Trinitron. It was also rectangular, and the size was about 13 in (33 cm). It had an in-line gun assembly, and the phosphor screen was striped rather than in triads. The shadowmask was made of a grille with vertical slots running all the way from the top to the bottom.

Wide-angled tubes came along in 1970, with a deflection angle of 110°. This enabled the tube and the cabinet to have less depth.

Many new types of tube were developed in the late 1970s, such as tubes with magnetic shields inside the glass envelope, and tubes with quick-heat cathodes. The Japanese invented tubes with distortion-free pictures. Later on, tubes became very rectangular in shape. They also had very flat screens.

Philips introduced the 20AX 'Great Leap Forward' in about 1974. Computer-aided design allowed production of a self-converging yoke based on the mathematical predictions of their own Research Department in 1954. Philips also introduced a tube in 1984 called the 45AX. It also had a self-converging yoke with very few adjustments to ensure beam convergence.

The delta tube

Figure 21.1 shows the internal layout of a delta gun tube. The guns are offset radially so that there is an angle of 120° between any two. Approximately 1 cm behind the screen glass there is a shadowmask. It contains numerous holes: one hole for every group of three phosphor dots. When the beams are correctly aimed at a hole, they can 'see' only their own phosphor dots.

Tube manufacture

When the tube is made, the dots for any one phosphor are positioned so that only an electron beam from the correct gun and also the correct point of deflection will strike the dots correctly. The beam in question cannot 'see' the other phosphors because of the shadowmask; it prevents the wanted beam from striking the wrong phosphor. An incorrectly adjusted CRT, a fault in manufacture, or stray magnetic fields can cause the three beams to strike the wrong phosphors. If this happens, dots other than the correct ones will fluoresce (give off light). The raster will appear to have on it patches of colour where they clearly should not be, even on a monochrome picture. Careful manufacturing processes and adjustments normally ensure that the shadowmask CRT will give years of trouble-free service to the user.

Colour purity adjustment

For colour purity, each beam must be made to pass through its own colour deflection centre. If one of the

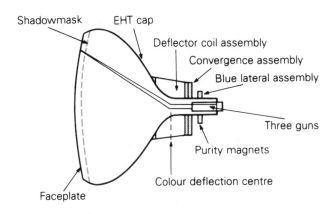

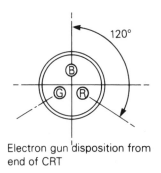

Electron gun disposition from end of CRT

Figure 21.1 The basic layout of a colour CRT (delta tube).

beams is made to do this (usually the red), then the other two beams will also be correct.

Colour purity adjustment is carried out in two stages. First, the three beams are aimed at the screen centre; second, they are made to bend at the correct place – that is, at their colour deflection centres.

To carry out the first task, the deflection coils are deliberately maladjusted by sliding them backwards or forwards, away from the colour deflection centre position. This will ensure that the screen will become 'impure'. If the red gun only is used, it can be aimed at the screen centre because it will fall only on the red phosphor dots in the undeflected position. The centre of the screen is the undeflected position. Elsewhere, it will fall on the phosphors creating blue and green patterns. In the centre it will fall only on its own phosphor and show up as a red ball.

The patch can be centralised by means of **purity magnets**, which operate in the same way as the shift rings on a monochrome CRT. When this adjustment has been done, the correct deflection centre for colour purity over the entire screen can be found by sliding the deflection coils along the tube axis until the purity is at its best. This adjustment ensures that the beams from the red, green and blue guns strike only their own phosphors within their own colour triad.

Convergence adjustments are performed in two stages: static convergence and dynamic convergence. The purpose of **static convergence** is to cause the beams from the red, green and blue guns to coincide at the screen centre. The word 'static' implies that there is no movement: that is, right at the screen centre, the point of no deflection or static area.

For this adjustment a pattern generator is needed which will give a 'crosshatch' pattern. Connect the generator to give a normal display. Now, concentrate on the centre crossing line. Switch off the blue gun.

Using the static convergence adjustments on the convergence assembly on the tube neck, shift the red permanent magnet. Two sets of grids should now be visible, one green and one red. The two separate centre crosses should now be at 30–40° to each other. Move the green static magnet so that the two crosses coincide with each other. When the red and the green grids coincide at the screen centre, this is the position for red and green static convergence. The colour of the centre of the cross should now be yellow.

Switch on the blue gun. Adjust the blue magnet so that the blue horizontal lines coincide with the yellow at the screen centre. The centre cross horizontal line should now be white. Adjust the blue lateral magnet to line up the blue vertical lines at the screen centre. There now should be a white crosshatch at the screen centre. If necessary, repeat the operation.

The purpose of **dynamic convergence** is to cause the beams from the red, green and blue guns to coincide in all the other areas away from the centre of the screen: that is, when the beams are moving. Perfect convergence is impossible with any type of tube. So these adjustments are carried out to achieve the best overall convergence over the entire face of the screen.

Many different methods are employed by different manufacturers to achieve good convergence. You should study receiver service manuals to gain experience.

Modern colour tubes

Modern colour tubes are of the **self-converging** type. The design is such that the three beams are kept together during their journey across and down the screen face.

Convergence correction is built into the design of the scanning coils. These coils give varying densities of magnetic fields in different parts of the deflection centre. This keeps the beams converged in all areas of the screen. For tubes of this type, the scanning coils are married to a particular tube and stay with the tube for the rest of its life.

Table 21.1 Typical voltages for the colour CRT

Heater	6.3 V, a.c. or d.c.
Cathode	90–160 V d.c.
Grid	0–60 V
First anode (a1)	400–800 V d.c.
Third anode (a3) focus	3–6 kV d.c.
Final anode (a2/a4)	22–30 kV d.c.

A change of tube therefore usually means a change of scanning coil assembly as well.

The result of this new technology in tube design is that a technician does not have to waste many hours performing convergence adjustments, as in the early days of colour television.

There are many new types of display tube on the market today. To explain the operation of every type would be beyond the scope of this book. It is left to the student/technician to keep up with all forms of new technology in this area.

Typical voltages for the colour tube

These may vary according to the size and the type of receiver chassis, but typical figures are given in Table 21.1.

REVISION EXERCISES AND QUESTIONS

1   A colour TV is displaying the following vertical colour bars. From left to right, the colours are: white, blue, magenta and yellow. Draw one picture line that would be visible on an oscilloscope at the red cathode. Include in your sketch the following:
   i)   the time duration for the active part of the line;
   ii)   the time duration of the blanking pulse;
   iii)   the approximate peak-to-peak voltage of this signal.

2   What would be the effect on the picture in each case for a colour television with the following faults:
   i)   The green gun has a short circuit between the grid and cathode.
   ii)   The red gun is suffering from zero emission.
   iii)   A fault causes the cathode of the blue gun to rise well above 190 V.

For all the above conditions, assume that the receiver is displaying normal colour bars.

Answers to questions and answering hints

Introduction

This section provides you with all the answers to the variety of questions and exercises given in the book. Always try a question or exercise yourself before you look at the answer. This will increase your understanding of the topic and give you practice in answering questions. If you are not sure of a particular answer, re-read the relevant section or chapter in the book to revise the work. You need to understand why a question has a particular answer, so that you can apply your understanding to similar types of question or exercise in your examinations and course assignments.

The book contains a variety of types of question and exercise. Find out the types of question that you will be expected to answer and their pattern. If possible, obtain past papers to support your work and revision. Some of the questions in the book require longer answers. We have provided hints on how to tackle these questions, and on the range of topics that you should include. Practise giving full answers to these questions and then check the answering hints to see that you have included all the relevant topics.

To revise a topic quickly you can also refer to the 'Check your understanding' sections given at the end of each chapter, and the list of key words with definitions given at the end of the book.

Hints to answering questions in examinations and course work

- Read all the questions carefully before you try anything. Make sure that you understand what each question is asking you to do.
- Plan the time that you will spend on each question. Use the marks as a guide: the more marks a question is worth, the more time it is worth spending on it.

- If you have a choice of questions, try to make your choice and stick to it. Don't change your mind halfway through the examination.
- Make sure that you earn all the 'easy' marks. Do not spend too long on a question you find difficult. Leave it; if you have time, you can try it again later when you have finished all the other questions.
- Keep an eye on the time. Make sure that you try all the questions you are required to answer.
- Always present your work as clearly as you can, whether you are writing or drawing. Make your work easy to follow for the examiner or assessor.
- Try and allow some time at the end to check your answers and improve them.
- In practical work, make sure that you understand what you are being asked to do by re-reading the question before you start. Follow all instructions carefully.

Chapter 1

1 500 Hz; harmonics are multiples of the fundamental.
2 Usually demonstrated using a diagram: see Figure 1.2.
3 Divide the wavelength into the speed of the EM wave through free space: that is, 300 000 000 m/s. i) 10 GHz; ii) 300 kHz; iii) 1500 kHz.
4 See 3, but use the speed of sound through air.
   i) 100; ii) 38.5.
5 See Figure 1.8.
6 The electric wave will now be horizontal and the magnetic wave vertical in the model: see text drawing.

Chapter 2

1 See text on the ionosphere and surface wave (ground wave propagation). At night the D layer virtually disappears as far as MF are concerned, therefore no absorption takes place in the D layer as happens during the day. MF can then have a sky-wave component, using the E layer to refract the wave back

to earth. This means that radio stations will be received from a greater distance away than during the day. It can also cause annoying interference when the sky wave reception interferes with the ground wave reception.

2 See text on the ionosphere and communications using the sky wave.

3 See ground wave propagation. The distance that the ground wave travels is dependent on the absorption that takes place in the earth's surface, which in turn depends on the terrain that the ground wave travels over.

4 Directional aerials are necessary to allow maximum transmission and reception to be achieved in a given direction. See text on transmission and reception of EM waves.

5 See text on communication using sky waves, fading and diversity reception.

6 Radio or EM waves induce voltages in the receiving aerials only if the polarisation is the same as that of the transmitter.

Chapter 3

1 See introduction text.

2 i) 50% modulation indicates that the modulating waveform is half the amplitude of the carrier waveform; therefore the diagram should indicate the carrier envelope rising and falling above and below the carrier waveform by half the carrier amplitude. See text on AM and percentage modulation.

   ii) See Figure 3.3a indicating 100% modulation and Figure 3.3b of over 100% modulation. The envelope of the AM does not follow the modulating waveform: therefore distortion of that waveform will occur when demodulation takes place.

3 i) A lower sideband range of frequencies: 15 996 000–15 999 650 Hz.

   ii) The carrier frequency of 16 MHz.

   iii) A higher sideband range of frequencies: 16 000 350–16 004 000 Hz.

4 See Figure 3.7.

5 See text on advantages/disadvantages of AM and angle modulation respectively.

6 See text on pulse modulation.

7 Binary (digital); see text on PCM.

8 This is associated with the amount of bandwidth that is required for FM: at the lower bands there is not enough space to allow a reasonable number of FM broadcasts to be transmitted.

Chapter 4

1 Using the formulae given in the text under frequency of oscillation, and new frequency of oscillation: i)

2906.132 Hz; ii) 2904.65 Hz. Therefore difference of frequency introduced by loss is 2906.132 − 2904.65 = 1.482 Hz.

2 See text and Figure 4.10.

3 See text under piezoelectric effect.

4 Master oscillators need to be temperature controlled, have a stable power supply and to work into a constant load; see text

5 See text: AM transmitter.

6 See text: AM transmitter.

7 See Figure 4.14 and text explanation.

8 See text: AF amplifier.

9 See text: varactor diode.

10 See text: need for power stages

11 Matching the output stage of the transmitter to the aerial so that maximum power transfer can take place: see text on output stages.

Chapter 5

1 See text on the developed TRF receiver and Figure 5.3.

2 As a fixed tuned receiver the TRF forms the basis of all receivers. Its disadvantages lie in when the TRF is made tunable: see text on the advantages and disadvantages of a TRF receiver.

3 This question refers to the signal frequency amplifier, frequency changer and local oscillator stages in more expensive receivers, while in broadcast receivers only the frequency changer and local oscillator (mixer) would be considered: see text on superhet receivers.

4 The main advantage that comes from using a signal frequency stage is that of an increased signal-to-noise ratio for the receiver. However, there are several other advantages gained: see text on superhet receivers and SF amplifiers.

5 See text: detector or demodulator.

6 See text: superheterodyne receiver.

7 The answer to this question needs to emphasise the fact that second or image channel occurs only if there is a station on the frequency that is twice the IF away from the wanted received frequency. See text: types of interference that may be introduced by the superheterodyne receiver.

8,9 See text as in question 7.

10 See text: tracking.

11 Local oscillator tuned to SF + IF = 2320 kHz. Image channel, if present, would be at SF + (2 × IF) = 2790 kHz.

Chapter 6

Questions in this chapter are mainly descriptive, and answers are all to be found in the text, which describes individual circuits in some detail.

Chapter 7

1 Using the text on the AM and FM receiver block diagram, construct a block diagram that uses common blocks and switches blocks that are associated with either AM or FM reception only. Differences between the two systems will be represented by the switched blocks.
2 Refer to text on demodulation in an FM receiver.
3 See text: advantages and disadvantages of FM over DSB AM.
4 This question follows on questions in the previous two chapters on the same subject: see text on signal frequency amplifiers.
5 Noise is the main consideration here, as in the previous question.
6 Refer to text: the section on slope detectors. In this case the AM detector is acting as a slope detector.
7 Refer to text on AFC circuits with reference to FM receivers.
8 See text: muting circuits.

Chapter 8

1 See text: frequency synthesis.
2 As question 1.
3 See text: stereo broadcasting. Stereo broadcasting must be able to be received by mono receivers as mono broadcasts.
4 See Figure 8.7. The 19 kHz transmission is transmitted as a pilot carrier transmission for the benefit of the receiver. Note that it can also be used to energise an indicator light at the receiver, showing that stereo broadcasts are being received.
5 This is the shape and scope of the reception pattern transmitted by the satellite on to the Earth's surface. See text.
6 See text on satellite communication: refer to the up and down link.
7 See text and Figure 8.5 in section on satellite communications.
8,9 See text on PLL circuits.

Chapter 9

1 The raster flicker rate for the British 625-line system is 50. Two fields make one picture.
2 For this answer you should have read the section on interlaced scanning. The flicker rate is increased to 50 by using interlaced scanning. Each field scan is broken down into 312.5 lines instead of 625 lines. This means that to produce one picture two complete fields are required, each one being of 312.5 lines.
3 Lost lines are unavoidable because of the beam's path during field flyback. Lines are wasted during this process. They would be in the wrong position on the screen.

4 Very simply, to synchronise the two timebases in the receiver. This will ensure that the scanning beam (receiver) will be in the correct place at the right time.
5 The ability of the human eye to retain an image on the retina after the light from that image has disappeared.
6 The aspect ratio is the ratio of the width of the screen to the height of the screen. For terrestrial television, the aspect ratio is 4:3.

Chapter 10

1 Negative vision modulation is preferred because of the effect on the picture of impulse interference. Black spots are less noticeable than white spots.
2 Using single-sideband transmission for television transmissions is a bit risky, in that it may be possible actually to lose the carrier. This would be mainly due to the problems of designing a filter that will filter very close to the carrier. A compromise is to use vestigial sideband transmission.
3 To provide impedance matching. It prevents the video amplifier from loading the vision demodulator.
4 15 625 scanning lines in every second ($25 \times 625 = 15\,625$).

Chapter 12

1 The purpose of weighting is to prevent overmodulation. The $R - Y$ and $B - Y$ signals are reduced in amplitude before being added together and then added to the luminance signal. They are then renamed V and U.
2 To assist separation in the receiver. Without this $90°$ shift, separation would not be possible.
3 $Y = (0.59G) + (0.30R) + (0.11B)$
4 White, yellow, cyan, green, magenta, red, blue and black.

Chapter 13

1 To synchronise the reference oscillator in the receiver. Also, it informs the receiver that a PAL or an NTSC chroma line is being received. It is also used for ACC purposes.
2 A way of transmitting extra information (such as the chrominance signal) within the sidebands of the luminance signal. The slotting-in of this extra information is known as interleaving or interlacing.
3 This deliberate distortion of the V chroma signal is done to counteract the effects of phase errors on the chroma signal, after it has been transmitted. The receiver can correct for this distortion and so cancel out these phase errors.
4 So that it can fit into the spaces of the luminance sidebands exactly. This will produce the least amount of interference on the screen of a monochrome receiver.

Chapter 14

1 To select, amplify and change the frequency of the transmission frequencies.
2 The surface acoustic wave filter.
3 If this did occur, the sound might be lost. The reason for this is that, in the receiver, intercarrier sound is used. This relies on having the two IF signals present. If the vision carrier were allowed to reduce to 0 per cent, then it might cancel out on peak white and disappear. The result of this is that the sound would be lost.
4 Less input signal required and the output video signal is of better definition because of the process used.
5 Impedance matching between the vision demodulator and the video amplifier.
6 The maximum voltage swing at the collector of the video output transistor cannot exceed the supply voltage. If the supply voltage was about 12 V, then this figure would represent the maximum swing available, i.e. 0–12 V. This swing of 12 V would not be big enough to drive the cathode voltage to obtain black or white on the screen. For a small-screen portable TV the cathode requires a voltage swing of about 80 V to get black or white on the screen (say from 50 V to about 130 V). The high HT required in a portable receiver is usually obtained from the line output stage. See the section on the line timebase.
7 The approximate size of the video signal at the output of the vision demodulator is 1–2 V peak to peak.
8 The gain of the IF stage is low at these two frequencies because 31.5 MHz and 41.5 MHz represent the unwanted adjacent vision and adjacent sound IFs respectively. These two signals are unwanted and are rejected.

Chapter 16

1 It is not perfectly linear because of the need for S correction. A perfectly linear scan waveform would produce distortions in the picture. This would be especially noticeable on round objects.
2 Crossover distortion is an effect that occurs in class B output stages. In the field output stage a class B stage is normally used. Crossover distortion occurs when one transistor 'appears' slow in taking over from the other, which has just turned off. When it finally turns on, the other transistor has long been off. Therefore a distortion occurs in the output waveform. This distortion can be overcome by biasing both transistors very slightly on.
3 S correction is achieved in the field timebase by using negative feedback.
4 To allow for the scanning of a nearly flat television screen. If S correction were not applied, the picture would appear distorted on some scenes, especially round objects. They would appear 'egg-shaped'.

Chapter 17

1 The advantage of using flywheel synchronisation is that it has superior performance when the transmitted signal is suffering from impulse interference. The flywheel sync circuit does not rely on receiving every line sync pulse that is transmitted, but rather a succession of pulses. The flywheel action produced by the circuit has what is known as 'momentum': the circuit will try to keep the oscillator running correctly.
2 The only disadvantage is that if the receiver is operated with a videotape recorder, the response time of the circuit tends to be a bit slow. The result of this is that picture distortion takes place. Modifications can be made to the receiver to make the picture tolerable to watch. See the section on the flywheel sync stage.

Chapter 18

1 The main advantage of using the intercarrier sound technique is that it allows for the drifting of the local oscillator in the tuner unit. This means that the sound IF will not be lost. See section on intercarrier sound.

Chapter 19

1 Remove the mains supply from the receiver and discharge the final anode capacitor.
2 Screen symptoms are as follows: dark picture, even when the brightness control is fully advanced (turned up); picture highlights (the bright parts) just visible.
3 i) When the receiver is turned on and warmed up, there is no effect on the picture.
    ii) When the receiver is turned off, a white spot should appear in the middle of the CRT screen.

Chapter 20

1 The main difference is that a simple PAL receiver does not have a PAL delay line.
2 In the case of simple PAL, phase errors are optically averaged out by the human eye. PAL D receivers electrically average out any phase errors by using a special delay line.
3 A notch filter.
3 The luminance amplifier is of the wideband type. Signals travel at a faster rate through this type of amplifier than signals travelling through narrow-band amplifiers. Therefore the luminance signal must be delayed in time so that it arrives at the CRT cathode at the same time as the chroma signals.
4 The luminance signal would be displaced to the left of the chroma signal by about 0.36 in (9.1 mm).

Chapter 21

1 For this answer, see Figure A.1.

2 i) For this fault symptom, a short-circuit between the grid and cathode would reduce or get rid of the grid/cathode bias. The visible symptom would be that the screen would appear very bright green. No information for red or blue would be noticeable due to this fault condition.

   ii) The red gun suffering from zero emission would produce, on a monochrome scene, a cyan-tinged picture. For the colour bars or any colour picture, reds, yellows and magentas would be affected.

   iii) If the blue gun/cathode were to rise well above 190 V, this excessive voltage would be enough to turn off the blue gun. Therefore a monochrome scene would appear to have different shades of yellow. For colour bars, white would appear as yellow. The yellow bar would not be affected. The cyan bar would appear as green. The green bar would be unaffected. The magenta bar would appear as red. The red bar unaffected. The blue bar would appear as black. Finally, black would be black.

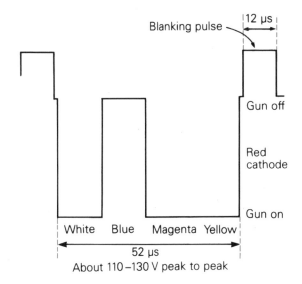

Figure A.1 Chapter 21, question 1.

Key words and definitions

ACC automatic chrominance control.

accepter circuit Tuned circuit responding to a signal of one specific frequency.

acuity The sharpness of our vision; the ability of our eyes to see small objects.

additive mixing Process in which two frequencies are added together, and then applied in series to a square-law device such as a diode or transistor circuit.

adjacent channel IF Unwanted IF generated in a TV receiver under extreme reception conditions.

adjacent channel interference Interference that is near the required frequency of reception.

adjacent channels The two channels either side of the wanted channel.

aerial diversity A system of **diversity reception** that accepts the strongest signal from two receiving aerials and their associated receivers, spaced well apart.

aerial Transducer used for the reception or transmission of radio signals.

AF amplifier An amplifier used for amplifying audio frequency signals.

AFC Automatic frequency control; used in radio and television receivers to keep the oscillator frequency constant.

AGC *See* **automatic gain control**.

alternating current Electrical current whose flow alternates in direction, flowing one way for one half cycle and reversing in direction for the next.

amplitude modulation Type of modulation in which the amplitude of a high-frequency carrier wave is increased or decreased by the amount of the amplitude of a low-frequency wave. Abbreviation AM.

amplitude The maximum value of the excursion of a wave, in a positive or negative direction.

angle modulation Type of modulation in which the phase angle of a constant carrier frequency is altered. Includes **frequency modulation** and **phase modulation**.

anode In a valve or tube, electrode held at positive potential with respect to the cathode, and through which positive current normally enters, by collection of electrons.

aquadag Graphite coating on the inside of a CRT, for collecting secondary electrons emitted by the face of the tube.

aspect ratio The ratio of the viewable width to the viewable height of a picture on a CRT.

automatic gain control Control circuit that maintains a constant output signal despite variations in input signal strength.

back porch In a TV signal, the portion of the synchronising pulse that immediately follows the line synchronising pulse.

balanced modulator A modulator that produces only the sidebands and not the carrier.

ballooning Visible effect created by an unstablised EHT system in a television receiver or monitor. The picture size appears to change in size depending on picture content.

bandpass coupling In an IF amplifier, technique in which a number of tuned circuits are coupled together using RF transformers.

bandwidth The spread of frequencies taken up by a specific channel or a specific type of signal.

beam limiter Circuit in a television receiver which prevents the beam current in the CRT from being exceeded.

blanking Dedicated function that prevents the beam in a CRT from being seen during line and field flyback.

buffer Intermediate stage, usually an emitter follower, which isolates one stage from another.

charged-coupled device Solid-state device used in modern TV cameras, which has replaced completely the old type of vidicon tube. Abbreviation CCD.

capacitor Electronic component capable of storing a charge of electricity. Can be used for smoothing, or for passing signals from one stage to another.

capture effect Suppression of a weak FM signal by a strong FM signal on the same or nearly the same frequency.

carrier frequency Output of a transmitter, onto which the required signal is modulated.

cathode modulation Modulation of the beam current of a CRT by feeding the video signal to the cathode.

cathode ray tube Electronic tube in which a stream of electrons is directed onto a surface to give a visible effect. Abbreviation CRT.

chromaticity The properties of the hue and saturation of a particular colour.

chromaticity diagram Diagram in which one of the three chromaticity coordinates is plotted against another.

chrominance signal The part of the TV signal that carries colour information. Also called the chroma signal.

co-channel interference Interference from a broadcasting station using the same frequency as the wanted station.

coincidence detector Type of FM detector used in a TV receiver to demodulate the **intercarrier sound** signal.

colour bar display Test pattern transmitted by broadcasters, used for checking the performance of colour television receivers.

colour burst Special sync signal transmitted with the composite video signal. Its prime purpose is to synchronise the reference oscillator in the colour receiver.

colour difference signal Signal produced by subtracting the Y signal from the camera outputs. In the receiver, it is added back to the luminance signal to produce the CRT drive signal.

colour killer circuit Circuit in a colour receiver designed to turn off the chroma stages when a monochrome transmission is received.

colour triangle Triangle drawn on a **chromaticity diagram**, showing the range of colours obtainable by additive mixing of three primary colours represented by the corners of the triangle.

compatibility The ability of a monochrome receiver to receive a colour transmission, and show a monochrome picture of reasonable quality.

contrast The ratio of white to black in a television picture. Adjusted by use of the contrast control.

cross colour Effect produced on a colour receiver screen when the picture contains luminance information corresponding closely to the frequency of the chrominance subcarrier. It produces fine patterning on the screen in the area of this luminance.

cross luminance In a colour receiver, slight incompatibility created on a monochrome picture by inclusion of the chrominance signal. In areas of saturated colours incorrect grey scale will result.

CRT *See* **cathode ray tube**.

crystal set Early radio receiver, with a crystal detector stage for demodulating the received signal, but no amplifier stages.

crystal-controlled oscillator An **oscillator** that contains a quartz crystal rather than a **parallel-tuned circuit**.

damped oscillation Output from an oscillator circuit when there is no positive feedback, starting at maximum and dying away to zero.

dark current Small leakage current flowing through a photoconductive material even when the material is in complete darkness.

decibel Unit used to express a ratio of power, voltage, or current. It can represent a loss or a gain.

de-emphasis In FM reception, the selective boosting of low audio frequencies to reverse the **pre-emphasis** that was applied before transmission.

demodulator Special circuit within a radio or television receiver that recovers the original modulation: i.e. the part of the transmission that contains the required information.

deviation In **frequency modulation**, the amount by which the carrier frequency is varied either side of its unmodulated position.

dichroic mirror Colour-selective mirror used within a colour television camera, which reflects a particular band of spectral energy and transmits all others.

differential phase distortion Unwanted phase changes of the chroma signal occurring somewhere between the studio camera and the circuitry within the colour receiver, resulting in changes of the colour on the screen.

diversity reception System designed to reduce **fading**. There are two types: **aerial diversity** and **frequency diversity**.

dot crawl Effect created on the screen of a monochrome receiver by the colour subcarrier.

double-sideband transmission Transmission of both the **sidebands** that are produced by conventional **amplitude modulation**.

downlink In satellite communications, the path consisting of the satellite transmitter output, the transmitting aerial, and the area on the earth (the **footprint**) served by the transmission.

electromagnetic wave Wave comprising an electric wave and a magnetic wave, 90° apart, which travel together through free space at the speed of light (300 000 000 m/s).

emitter follower Circuit with high input impedance and low output impedance.

envelope The outline of an RF carrier or IF signal complete with the modulation.

fading Variation of strength of the signal from a distant transmitter.

field blanking Period between successive TV fields, during which picture information is suppressed and sync pulses are transmitted.

field flyback In a CRT, the action of the electron beam moving from the bottom of the screen to the top.

field timebase In a CRT, the circuitry that causes the electron beam to move in the vertical direction. It consists of the field oscillator and the output stage complete with scan coils.

flicker rate The number of complete fields in one second.

footprint In satellite communications, the area on the earth's surface within which a satellite broadcast gives satisfactory results.

frequency changer Another name for the **mixer** circuit in a **superheterodyne** receiver.

frequency diversity System of **diversity reception** that accepts the strongest signal from two frequencies of a common signal.

frequency modulation Type of modulation that uses a carrier of constant amplitude; the instantaneous frequency of the carrier is varied in sympathy with the modulating frequency.

frequency swing *See* **deviation**.

frequency synthesis Production of signals of precisely defined frequency from one or more **crystal-controlled oscillators** by means of multipliers, dividers and mixers.

frequency The number of times a wave completes a cycle of values in 1 second.

front porch Part of a TV synchronising pulse; comprises a short interval of black level immediately before the line synchronising pulse.

gamma correction System of signal correction in a television transmitter to overcome inherent non-linearity within the camera tube and the receiver CRT.

ganging Mechanical or electronic coupling of the oscillator and aerial circuits in a superhet receiver; more generally, mechanical coupling of variable circuit elements (such as capacitors) for simultaneous control.

geostationary orbit The orbit of a satellite that has an orbit time of 24 hours, equal to the rotation of the earth, and therefore appears to be stationary in the sky.

grid modulation Modulation of the beam current of a CRT by feeding the video signal to the grid electrode. Less efficient than **cathode modulation**.

ground waves Radio waves reflected from the ground.

Hanover bars Effect on the screen of a colour TV receiver when the chroma signal is suffering from phase errors. Also known as venetian blind effect.

harmonic Waveform whose frequency is a multiple of the frequency of a fundamental sine wave. A square wave is rich in odd harmonics: 3, 5, 7, 9, ... etc. A triangular wave is rich in even harmonics: 2, 4, 6, 8, ... etc.

hertz Unit of frequency; the number of complete cycles in one second. It can apply to an electromagnetic wave or to a signal flowing inside a conductor.

HF compensation Method used to boost the high frequencies relative to the low frequencies in a system.

hue The perception of colour that discriminates between different colours as a result of their wavelength.

ident signal Signal developed within a colour television receiver, which identifies a normal chroma line or a PAL chroma line.

IF *See* **intermediate frequency**.

illuminant D Reference white used as a colour television standard. Obtainable by mixing the correct portions of red, green and blue at the receiver.

image channel In a **mixer** (frequency changer), an unwanted signal frequency spaced from the wanted frequency by twice the IF. In the superhet receiver, the signal frequency stage is made selective in order to achieve adequate image channel rejection.

impedance Complex sum of a.c. reactance and resistance.

impulse noise Spikes of voltage caused by sparks when electricity is turned on and off.

intercarrier sound System of sound IF used in colour TV receivers. The sound signal is frequency modulated and transmitted on a signal 6 MHz higher than the video signal carrier.

interlaced scanning Scanning system in which the lines of successive fields are displaced to form an interlaced pattern of alternate lines.

intermediate frequency The frequency used for the main amplifying stages in a **superheterodyne** receiver. Abbreviation IF.

ionosphere The wide belt of ionised gases that surround the earth.

isotropic radiator An idealised theoretical aerial that sends out energy equally in all directions from a point source.

keying pulses Special pulses used in a PAL encoder to develop the burst signal.

limiter Circuit that prevents the amplitude of a signal from exceeding a preset level.

line flyback The part of a TV line timebase waveform in which the voltage or current returns to its starting value.

line timebase In a CRT, the circuitry that causes the electron beam to move in the horizontal direction.

line-of-sight communications Point-to-point communications achieved when the receiving aerial is in sight of the transmitting aerial.

local oscillator The oscillator within a **superheterodyne** receiver. The signal from the LO beats with the incoming signal to form the **intermediate frequency**.

losses Any deficiencies in any electronic system that attenuate signals at certain or all frequencies. Losses are unwanted in most systems. They can be put to use: e.g. eddy current losses are used to prevent damage to the movement in a moving coil meter.

lower sideband *See* **sideband**.

luminance delay Delay applied to the luminance signal in a PAL encoder or decoder.

luminance signal The part of a TV signal that carries brightness information.

mixer modulation The process of impressing onto an RF carrier signal the wanted modulation or intelligence.

mixer Circuit that produces the sum and difference of two input frequencies. Used in the **superheterodyne** receiver.

modulation index In **frequency modulation**, the ratio of the frequency **deviation** to the frequency of the modulating signal.

Morse code Old system used for sending messages, in which each letter of the alphabet is sent as a combination of dots and dashes.

multiplicative mixing Process in which two frequencies are multiplied together in an FET or dual-gate MOSFET.

multivibrator Form of oscillator circuit used for producing non-sine waves.

negative feedback Process in which part of an amplifier's output is fed back to the input signal so as to reduce the total input signal.

negative vision modulation Inversion of the video signal before modulating the RF carrier. Used to reduce the effect of impulse interference on the video signal and hence the visible effect on the screen.

neutralisation Technique used in tuned amplifiers (especially valve circuits) to prevent instability caused by internal positive feedback.

neutralising capacitance Adjustable capacitor intended to cancel out the internal capacitance between grid and anode of a high-frequency valve amplifier, to prevent possible instability caused by feedback from output to input.

Nipkow disc Disc with holes arranged in a spiral; used in early television systems to provide a form of mechanical scanning.

NTSC National Television Systems Committee.

orthogonal scanning System in which the path of the electron beam within a television camera tube is at right angles to the target layer, to reduce the effects of distortion.

oscillator Electronic circuit that produces an a.c. signal from a d.c. source. There are three requirements for an oscillator: frequency determination, gain, and positive feedback from output to input.

PAL Phase alternation line.

parallel-tuned circuit Resonant circuit consisting of an inductor connected in parallel with a capacitor.

percentage modulation In **amplitude modulation**, the amount by which the amplitude of the carrier frequency is varied by the modulating signal.

persistence of vision Brief retention of an image on the retina of the eye after the light from an image has ceased; reduces the effect of flicker in viewing television and cinema.

persistence, CRT Emission of light from a CRT screen when the electron beam has been turned off, or has left that particular area.

phase modulation Type of modulation that uses a carrier of constant amplitude; the instantaneous phase relationship of the carrier is varied in sympathy with the modulating frequency.

phase shift keying Transmission of coded data using **phase modulation** with several discrete phase angles.

phase-locked loop Control loop comprising a voltage-controlled oscillator and phase detector to lock a signal to a stable reference frequency. Abbreviation PLL.

photoconductive layer Chemical layer whose resistance varies with the amount of light falling on it. Used in television camera tubes.

piezoelectric effect Effect in crystals such as quartz, which produces a voltage when the crystal is stressed, and changes the crystal's dimensions when a voltage is applied to it.

polarisation Non-random orientation of the electric and magnetic fields of an electromagnetic wave.

positive feedback Process in which part of an amplifier's output is fed back to the input so as to reinforce it.

preamplifier Amplifier that boosts the output of a low-level AF or RF source to an intermediate level for further processing without significant degradation of the signal-to-noise ratio.

pre-emphasis   Selective boosting of high audio frequencies in FM transmission to prevent their being submerged by noise in the system. On reception, the process is reversed by **de-emphasis**.

pulse modulation   Form of modulation in which the carrier is a stream of pulses of electrical energy rather than a signal of constant amplitude and frequency.

pulse-code modulation   Form of modulation based on the transmission of digital signals.

quadrature amplitude modulation   Modulation system involving phase and amplitude modulation of a carrier.

quantisation   Sampling of a signal to produce a number of values per second.

raster   The scanning pattern on a television screen.

ratio detector   Type of FM demodulator used in older colour TV receivers to demodulate the **intercarrier sound** signal.

receiver   Any piece of equipment capable of receiving a signal: for example, a radio or a television.

relative luminosity curve   Frequency response curve of the average human eye.

reverse compatibility   The ability of a colour TV receiver to display a satisfactory monochrome picture when tuned to a monochrome transmission.

rotating vector   Graphical method for plotting a sine wave.

saturation   The amount of purity of a colour.

scanning   The movement of the electron beam within the CRT. envelope.

SECAM   Séquentiel couleur à mémoire.

selective fading   Reduction in strength of parts of a radio signal that take different routes through the ionosphere.

selectivity   The ability of a radio receiver to select the wanted signal and reject the unwanted one.

shadowmask CRT   Original type of colour TV tube, containing perforations in groups of three to ensure that electrons from three separate electron guns (red, green and blue) strike the correct phosphor dots on the screen.

sidebands   Bands on either side of the **carrier frequency**, introduced by the process of modulation. In **amplitude modulation**, the upper and lower sidebands are equal to the sum and difference of the carrier and modulating frequencies respectively.

signal frequency amplifier   The RF amplifier in a **superhet** receiver.

sine wave   Waveform whose shape is that of a graph of the sine of an angle plotted against the angle. Describes the voltage produced by a coil of wire rotating in a uniform magnetic field.

skip distance   Distance between the end of the ground wave and the first return of the sky wave.

sky wave   Radio wave refracted through the ionosphere between transmitter and receiver.

slope detector   Simple FM demodulator, consisting of a single parallel-tuned circuit tuned away from the IF by a small amount.

space wave   Radio wave propagated directly from transmitter to receiver.

square wave   Pulsed wave with very rapid rise and fall times, and pulse duration equal to half the period of repetition. Can be shown to consist of a sine wave fundamental plus an infinite number of odd harmonics.

squelch circuit   Circuit used in an FM receiver to eliminate noise when the receiver is not tuned to a station.

stability (of an oscillator)   Oscillator's ability to remain stable on the required frequency.

stereophonic   In sound reproduction, the use of more than one channel to create the effect of an extended source of sound.

stratosphere   Layer of the atmosphere above the **troposphere**, in which temperature remains constant.

stray capacitance   Any capacitance occurring in a circuit other than that intentionally inserted by capacitors.

subcarrier   Frequency modulated over a narrow range by a measured quantity, and then used to modulate a carrier that will finally be demodulated on reception.

superheterodyne principle   The mixing together of one or more received signal frequencies with that of a locally generated oscillator signal at frequencies above the audible range.

surface wave   *See* **ground wave**.

teletext   Method of transmitting alphanumeric information to domestic TV receivers during field flyback.

time division multiplexing   Form of multiple transmission on a common frequency or channel by interleaving pulse-modulated signals.

timebase   Line formed by applying a particular waveform to a CRT scanning system. The **line timebase** causes horizontal movement of the beam; the **field timebase** causes vertical movement of the beam.

tracking   In a **superheterodyne** receiver, maintenance of the correct frequency difference between the RF stage and the **local oscillator** frequency.

transducer   Any device that converts one form of energy into another. Examples are a loudspeaker, a microphone, and an aerial.

transmitter   Apparatus used to produce and modulate RF current.

trap circuit   Acceptor or rejecter circuit that removes any unwanted signals.

troposphere   Layer of the atmosphere extending upwards from about 6 miles (10 km) above the earth's surface, in which temperature decreases with increase in height.

tuned radio frequency receiver   Radio receiver comprising a number of amplifier stages tuned to resonate at the carrier frequency of the desired signal.

tuner   The portion of a receiver that comprises the RF, IF and demodulator stages.

uplink   In satellite communications, the path consisting of the ground station, the transmission path and the satellite's receiving aerial.

upper sideband   *See* **sideband**.

varactor diode   Semiconductor diode with voltage-variable capacitance.

varicap diode   Variable-capacitance diode: a diode designed specifically for varying the resonant frequency of a tuned circuit by the application of a d.c. voltage.

venetian blind effect   *See* **Hanover bars**.

vestigial sideband transmission   Partial removal of one of the **sidebands** in a transmission system.

vidicon   One type of thermionic camera tube.

wavelength   The distance between corresponding points in a sound wave or electromagnetic wave: the shorter the wavelength, the higher the frequency.

white noise   Random electronic noise across the frequency spectrum.

Appendix 1
World television standards

Country	System	Colour	Number of sets	Country	System	Colour	Number of sets
Afghanistan	B	PAL	100 000	Canary Islands	B, G	PAL	240 000
Alaska (US State)	M	NTSC	N/A	Central African Republic	K	SECAM	7 500
Albania	B, G	PAL	3 000 000	Chad	D	SECAM	N/A
Algeria	B	PAL	2 000 000	Chile	M	NTSC	2 000 000
Angola	I	PAL	51 000	China (People's Republic)	D	PAL	228 000 000
Antarctica	M	NTSC	N/A				
Antigua and Barbuda	M	NTSC	28 000	Colombia	M	NTSC	5 500 000
Argentina	N	PAL	7 170 000	Congo	K	SECAM	8 500
Armenia	D, K	SECAM	N/A	Cook Islands	B	PAL	3 500
Aruba	M	NTSC	19 000	Costa Rica	M	NTSC	340 000
Ascension	I	PAL	N/A	Cuba	M	NTSC	2 500 000
Australia	B	PAL	8 000 000	Croatia	B, H	PAL	750 000
Austria	B, G	PAL	2 700 000	Cyprus	B, G	PAL	234 500
Azerbaijan	D, K	SECAM	N/A	Czech Republic	D, K	SECAM	N/A
Azores	B	PAL	3 400	Denmark	B	PAL	2 700 000
Azores (US Forces)	M	NTSC	N/A	Diego Garcia	M	NTSC	N/A
				Djibouti	K	SECAM	17 000
Bahamas	M	NTSC	60 000	Dominica	M	NTSC	5 200
Bahrain	B, G	PAL	270 000	Dominican Republic	M	NTSC	728 000
Bangladesh	B	PAL	350 000				
Barbados	M	NTSC	69 350	Dubai	B, G	PAL	N/A
Belgium	B, H	PAL	4 200 000	Ecuador	M	NTSC	900 000
Belize	M	NTSC	27 050	Easter Island	B	PAL	N/A
Belorussia	D, K	SECAM	N/A	Egypt	B, G	PAL	5 000 000
Benin	K	SECAM	20 000	El Salvador	M	NTSC	500 700
Bermuda	M	NTSC	30 000	Equatorial Guinea	B	SECAM	2 500
Bolivia	M, N	NTSC	50 000	Estonia	B/D, K	PAL/SECAM	600 000
Bosnia and Herzegovina	B, H	PAL	N/A	Ethiopia	B	PAL	100 000
				Falkland Islands	I	PAL	N/A
Botswana	K	SECAM	14 000	Fernando Po	B	PAL	N/A
Brazil	M	PAL	30 000 000	Faroe Islands	B, G	PAL	14 000
British Indian Ocean Territory	M	NTSC	650	Fiji	M	NTSC	N/A
				Finland	B, G	PAL	1 900 000
Brunei	B	PAL	70 000	France	L	SECAM	29 300 000
Bulgaria	D, K	SECAM	3 100 000	French Polynesia	K1	SECAM	N/A
Burkina	K	SECAM	46 000	Gabon	K	SECAM	40 000
Burundi	K	SECAM	4 500	Galapagos Islands	M	NTSC	4 000
Cambodia	B, G	PAL	70 000	Gambia	B	PAL	N/A
Cameroon	B	PAL	15 000	Georgia	D, K	SECAM	N/A
Canada	M	NTSC	17 400 000	Germany	B	PAL	30 500 000

Country	System	Colour	Number of sets	Country	System	Colour	Number of sets
Ghana	B	PAL	N/A	Mayotte	K	SECAM	3 500
Gibraltar	B, G	PAL	7 500	Mexico	M	NTSC	56 000 000
Greece	B, G	SECAM	2 300 000	Micronesia	M	NTSC	7 000
Greenland	B	PAL	21 000	Monaco	L, G	SECAM/PAL	20 000
Grenada	M	NTSC	30 000	Montserrat	M	NTSC	1 625
Guadeloupe	K	SECAM	150 000	Morocco	B	SECAM	1 200 000
Guam (US Territory)	M	NTSC	75 000	Mozambique	B	PAL	35 000
				Myanmar (Union of)	M	NTSC	1 000 000
Guatemala	M	NTSC	475 000	Namibia	I	PAL	38 500
Guiane	K	SECAM	6 500	Nepal	B	PAL	250 000
Guinea	K	PAL	65 000	Netherlands	B, G	PAL	6 500 000
Guyana	M	NTSC	15 000	Netherlands Antilles	M	NTSC	35 000
Haiti	M	NTSC	25 000				
Hawaii (US State)	M	NTSC	553 000	New Caledonia	K	SECAM	35 500
Honduras	M	NTSC	160 000	New Zealand	B	PAL	1 100 000
Hong Kong	I	PAL	1 750 000	Nicaragua	M	NTSC	210 000
Hungary	D, K	PAL	4 300 000	Niger	K	SECAM	25 000
Iceland	B, G	PAL	76 000	Nigeria	B	PAL	6 100 000
India	B	PAL	20 000 000	Norfolk Island	B	PAL	900
Indonesia	B	PAL	11 000 000	North Mariana Islands	M	NTSC	4 100
Iran	B	SECAM	7 000 000				
Iraq	B	SECAM	1 000 000	Norway	B, G	PAL	2 000 000
Ireland	I	PAL	1 000 000	Oman	B, G	PAL	1 500 000
Israel	B, G	PAL	1 500 000	Pakistan	B	PAL	2 100 000
Italy	B, G	PAL	17 000 000	Palau	M	NTSC	1 600
Ivory Coast	K	SECAM	810 000	Panama	M	NTSC	205 000
Jamaica	M	NTSC	484 000	Papua New Guinea	B, G	PAL	10 000
Japan	M	NTSC	100 000 000				
Jordan	B	PAL	250 000	Paraguay	N	PAL	350 000
Kampuchea	M	SECAM	N/A	Peru	M	NTSC	2 000 000
Kazakhstan	D, K	SECAM	N/A	Philippines	M	NTSC	7 000 000
Kenya	B	PAL	260 000	Poland	D, K	PAL	10 000 000
Korea (Republic)	M	NTSC	10 400 000	Polynesia (French)	K	SECAM	26 500
Korea (Democratic Peoples Rep)	D, K/M	PAL/NTSC	2 000 000	Portugal	B, G	PAL	1 700 000
				Puerto Rico	M	NTSC	830 000
Kuwait	B, G	PAL	800 000	Qatar	B	PAL	250 500
Kyrgyzstan	D, K	SECAM	N/A	Réunion	K	SECAM	90 500
Laos	B	PAL	80 000	Romania	D, G	PAL	4 000 000
Latvia	D, K	SECAM	1 200 000	Russia	D, K	SECAM	N/A
Lebanon	B, G	SECAM	1 100 000	Rwanda	K	SECAM	N/A
Lesotho	I	PAL	50 000	Samoa (American)	M	NTSC	8 000
Liberia	B	PAL	45 000	Samoa (Western)	B	PAL	5 000
Libya	B	PAL	500 000	San Marino	B, G	PAL	N/A
Liechtenstein	B, G	PAL	N/A	Sao Tome and Principe	B, G	PAL	21 000
Lithuania	D, K	SECAM	1 400 000				
Luxembourg	B, L, G	PAL/SECAM	101 000	Saudi Arabia	B, G	SECAM/PAL	4 700 000
Macao	I	PAL	70 300	Senegal	K	SECAM	61 000
Macedonia	B, H	PAL	N/A	Seychelles	B	PAL	13 000
Madagascar	K	SECAM	130 000	Singapore	B	PAL	650 000
Madeira	B	PAL	81 000	Slovakia	D, K	SECAM	N/A
Malawi	B, G	PAL	N/A		B	PAL	N/A
Malaysia	B	PAL	2 000 000	Slovenia	B, H	PAL	N/A
Maldives	B	PAL	4 750	Sierra Leone	B	PAL	25 000
Moldova	D, K	SECAM	N/A	Somalia	B	PAL	N/A
Mongolia	D	SECAM	135 000	South Africa	I	PAL	3 400 000
Mali	K	SECAM	10 000	Spain	B, G	PAL	17 000 000
Malta	B	PAL	133 000	St Kitts and Nevis	M	NTSC	9 500
Marshall Islands	M	NTSC	N/A	St Lucia	M	NTSC	25 000
Martinique	K	SECAM	65 000	Saint Pierre and Miquelon	K	SECAM	2 000
Mauretania	B	SECAM	1 100				
Mauritius	B	SECAM	157 000				

Country	System	Colour	Number of sets	Country	System	Colour	Number of sets
Sri Lanka	B	PAL	700 000	Ukraine	D, K	SECAM	N/A
St Vincent	M	NTSC	17 700	United Kingdom	I	PAL	20 000 000
Sudan	B	PAL	250 000	United States of America	M	NTSC	215 000 000
Surinam	M	NTSC	43 000				
Swaziland	B, G	PAL	12 500	Uruguay	N	PAL	600 000
Sweden	B, G	PAL	3 750 000	Uzbekistan	D, K	SECAM	N/A
Switzerland	B, G	PAL	2 500 000	Venezuela	M	NTSC	37 500 000
Syria	B	SECAM	700 000	Vietnam	M	NTSC/ SECAM	2 500 000
Tahiti	K	SECAM	N/A				
Taiwan (Republic of China)	M	NTSC	7 000 000	Virgin Islands (British)	M	NTSC	3 000
Tajikistan	D, K	SECAM	N/A	Virgin Islands (USA)	M	NTSC	31 500
Tanzania	B	PAL	80 000				
Thailand	B, M	PAL	3 300 000	Wallis & Futuna	K	SECAM	N/A
Tibet	D	SECAM	N/A	Yemen	B	PAL/NTSC	100 000
Togo	K	SECAM	150 000	Yugoslavia	B, G	PAL	1 600 000
Tonga	M	NTSC	2 500	Zaire	K	SECAM	22 000
Trinidad and Tobago	M	NTSC	250 000	Zambia	B	PAL	200 000
				Zanzibar	I	PAL	N/A
Tunisia	B, G	SECAM	650 000	Zimbabwe	B	PAL	137 000
Turkey	B	PAL	10 500 000				
Turkmenistan	D, K	SECAM	N/A				
Uganda	B	PAL	115 000				
United Arab Emirates	B, G	PAL	170 000				

Reproduced with permission from the 1995 edition of *The World Radio and TV Handbook*, copyright Billboard Books. (N/A = figures not available.)

Appendix 2
Test card F

Appendix 3
Trapezium distortion;
IF response curve

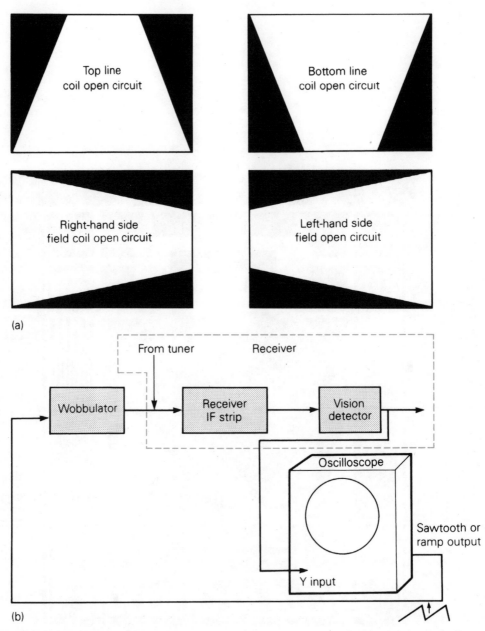

Figure A.3 (a) Four types of **trapezium distortion**, looking from the front of the screen. (b) Demonstration using a wobbulator or sweep generator to show the IF response curve of a 625-line system I television receiver.

Appendix 4
Units and abbreviations

SI base units and supplementary units

Quantity	Name of base unit	Unit symbol
Length or distance	metre	m
Mass	kilogram	kg
Time	second	s
Electric current	ampere	A
Thermodynamic temperature	kelvin	K
Amount of substance	mole	mol
Luminous intensity	candela	cd
Plane angle	radian	rad
Solid angle	steradian	sr

SI derived units

Quantity	Unit name	SI units	Unit symbol
Angular velocity	radians per second	rad/s	
Capacitance	farad	C/V	F
Conductance	siemens	A/V	S
Electric charge, electric flux	coulomb	A s	C
Electric potential	volt	J/C, W/A	V
Energy	joule	N m	J
Force	newton	$kg\,m/s^2$	N
Frequency	hertz	s^{-1}	Hz
Impedance	ohm	V/A	Ω
Inductance	henry	Wb/A	H
Luminous flux	lumen	cd sr	lm
Magnetic flux	weber	V s	Wb
Magnetic flux density	tesla	$Wb/m^2 T$	T
Power	watt	J/s	W
Pressure	pascal	N/m^2	Pa
Reactance	ohm	V/A	Ω
Resistance	ohm	V/A	Ω
Torque	newton metre	N m	
Velocity or speed	metres per second	m/s	
Weight	newton	N	
Work	joule	N m	J

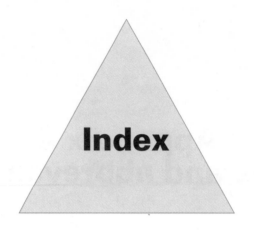

Index